P9-AFP-056

ALDRIN, DIELDRIN, ENDRIN and TELODRIN

AN EPIDEMIOLOGICAL AND TOXICOLOGICAL STUDY
OF LONG-TERM OCCUPATIONAL EXPOSURE

ALDRIN, DIELDRIN, ENDRIN AND TELODRIN

AN EPIDEMIOLOGICAL AND TOXICOLOGICAL STUDY OF LONG-TERM OCCUPATIONAL EXPOSURE

K . W . JAGER

*Industrial Medical Department of Shell Nederland Raffinaderij N.V.
and Shell Nederland Chemie N.V. Rotterdam*

ELSEVIER PUBLISHING COMPANY

AMSTERDAM / LONDON / NEW YORK

1970

ELSEVIER PUBLISHING COMPANY
335 Jan van Galenstraat
P.O. Box 211, Amsterdam, The Netherlands

ELSEVIER PUBLISHING CO. LTD.
Barking, Essex, England

AMERICAN ELSEVIER PUBLISHING COMPANY, INC.
52 Vanderbilt Avenue
New York, New York 10017

Library of Congress Card Number: 70 - 135487
ISBN: 0 - 444 - 40898 - 3
With 25 illustrations and 55 tables.

Printed in The Netherlands

From the Industrial Medical Department of
Shell Nederland Raffinaderij N.V. and
Shell Nederland Chemie N.V.
Rotterdam

Most laboratory data on which this study is based were determined in our
Clinical and biochemical laboratory, Head Miss A.C.M. Geeratz
and the
Chemical laboratory, Head Mr. J.G. Huisman

ACKNOWLEDGEMENT

This is to express my sincere appreciation to all those, who in one way or another stimulated, advised or assisted me in this study and in the preparation of this book, and to all insecticide workers for their splendid co-operation.

CONTENTS

A. General introduction

B. Survey of literature

C. Local situation

D. Results of study of long-term occupational exposure group

E. Results of study of all insecticide workers

F. Discussion and summary

A. GENERAL INTRODUCTION

Chapter 1

Introduction and scope of this study

1.1   USE OF PESTICIDES

1.1.1 Pesticides in general

Ever since the introduction of DDT as an insecticide in 1941, a large number of organic compounds have been synthesized and introduced as effective pesticides.

At the present time pesticides are considered to be indispensable for the production of an adequate food-supply for an increasing world population and for the control of insectborne diseases.

Before the widespread use of pesticides, insects, mites, nematodes, weeds and fungi caused crop losses which had such drastic effects on human life and the economy of the country concerned, that they have become part of world's history. The Irish potato famine of 1845 was responsible for a million deaths and for the emigration of over a million inhabitants. The disease of French vines in the latter part of the 19th century apparently did more damage to that country's economy than the Franco-Prussian war. Many countries are still facing the threat of calamity from invasion by locust-swarms. The Naples typhus epidemic of 1943-1944 would have been far more devastating had there not been DDT. Again, are not the recurrent epidemics which decimated crusaders, armies and populations, an intrinsic part of history?

Many pesticides are, however, very toxic substances and some are persistent in character. For agricultural and public health purposes this persistence is, at times, essential (Wright 1970). In fact, rural malaria control, for example, would have been impossible without persistent insecticides. This is one of the main reasons why the organochlorine insecticides such as DDT and dieldrin, which possess this characteristic, have been amongst the most successful insecticides yet discovered. The need for frequent applications does not occur with these compounds.

On the other hand, the very property of persistence now suggests a complex problem of environmental hygiene inasmuch as these pesticides are now being detected in our biosphere. Minute, but not necessarily increasing, residues of these pesticides found in foodstuffs, water and air have aroused concern over their possible effects on human health. Pesticide residues in man stem mainly from ingestion with food, also from occupational absorption through dermal exposure and from occasional contamination in the home and garden.

Concern about these residues is understandable: few topics will produce a greater public response than the suggestion that mankind is slowly poisoning himself and the environment.

1.1.2 Groups of pesticides

Pesticides comprise all types of synthetic and naturally occurring chemicals and are used in man's fight against those organisms which are harmful to his health and food production. Some of the groups into which these toxicants fall are:

insecticides,
nematocides,
fungicides,
rodenticides,
herbicides.

Insecticides may be divided into three main subgroups:
1. Organochlorine insecticides: chlorinated hydrocarbon compounds of widely varying structure.
2. Organophosphate insecticides: again with a wide variation in structure, but the common principle being that all are organic esters of phosphoric, phosphonic, or thiophosphoric acid.
3. Carbamates.

The insecticides in this study: aldrin, dieldrin, endrin and "TELODRIN"* are organochlorine insecticides of the cyclodiene subgroup.

1.1.3 Benefits to public health

Public health vector-control or pest-eradication programs in many countries, often recommended, coördinated and/or sponsored by the WHO, have vastly reduced morbidity and mortality from diseases that once took a heavy toll of life, e.g. malaria, yellow fever, typhus etc. Again, control of such diseases as filariasis, Chagas disease, Oroya fever, onchocerciasis and Leishmaniasis should now be possible.

---

*"TELODRIN" is the Shell trade-mark of the technical insecticide isobenzan (see chapter 3.2). For ease of reading we will subsequently write Telodrin.

It is precisely because of their stability, persistence and broad-spectrum action, that compounds such as DDT and the chlorinated cyclodienes are often the insecticides of choice for the control of the above diseases.

In veterinary practice pesticides are used in the control of parasites that attack cattle and sheep, *e.g.* blowflies, ticks and mites.

An additional and indispensable contribution to the improvement of world public health is the increase in food supplies made possible to a large extent by the use of pesticides. The world population is expected to double by the end of this century when the daily increase will amount to 150,000 people a day. Even now, according to the FAO, 500 to 1,000 million people have less food than they need and up to half of them suffer from real hunger or malnutrition or both (Sukhatme 1964). If widespread starvation is to be avoided world food supplies must be trebled by the year 2000 and without the use of pesticides this would be impossible.

Besemer (1965), in Holland, formulated this as follows (translated quotation): "Whoever propagates the discontinuance of the use of these (pesticides), is in reality, advocating the return of those terrors of our ancestors: hunger and pestilence".

1.1.4 Benefits to agriculture and food storage

The use of pesticides has produced remarkable increases in crop yields. Insecticides like aldrin, dieldrin and endrin provide the standard treatment against weevils, termites, wireworms, ants, most insect pests of cotton and many others. These same insecticides play a major role in the control of the desert locust. It is in this way that these insecticides contribute so considerably to the economy of many developing countries.

For a farmer there are two main reasons for using insecticides: firstly to insure his crop against serious loss due to massive infestation, and secondly to increase yields by limiting insect-damage throughout the growing period.

The major use of organochlorine pesticides in protecting stored food involves the control of pests, such as cockroaches and ants, that may invade foodstuffs while in store. Not only does the use of insecticides lead to a direct saving of food, it also helps to keep the food in a better condition. The insecticides are not applied of course, directly on to the food that is to be consumed.

1.1.5 Hazards in the use of pesticides

The beneficial properties of these insecticides: toxicity, broad-spectrum action and persistence, also cause many of the problems associated with their use.

Some of these problems are:
a. environmental contamination;
b. hazards to wildlife from acute exposure, cumulation or food-chain build-up;
c. potential hazards to man.

Of these, only environmental contamination and hazards to man are related to the subject of this study and will be discussed here.

1.2 ENVIRONMENTAL CONTAMINATION BY PESTICIDES

As a result of their persistence some pesticides have been found in many parts of our environment. The fact that this statement can be made, is to a large extent due to the vastly improved analytical techniques, which make it possible to detect residues as small as parts per 10^{12} in the environment.

The significance of the presence of pesticides in the environment was evaluated by Decker (1967): "The magnitude of any specific pesticide residue that one may expect to find in a given ecological niche, be it plant, soil, air, water, or animal tissue, will, as in book-keeping, represent the balance left after withdrawals have been substracted from total deposits. Any attempt to predict the magnitude of such residues presupposes a knowledge of the frequency as well as the rates of use, and a precise evaluation of all factors that govern or regulate the rate of loss or disappearance of the chemical. The persistence of any specific pesticide residue will be determined by a combination of many basically independent yet often interrelated factors, such as the chemical characteristics of the compound (its components and molecular configuration), its ability to withstand radiation, hydrolysis, oxidation, and reduction, its solubility, its absorption and partition coefficients".

The essence of this problem may very well be expressed, as was done by Kehoe (1965), who in discussing the potential human hazard from lead pollution, stated that the problem of environmental contamination is not a question of whether there is more or less contamination *per se*, but of whether "the quantities are within the limits that do not tax, unduly, the physiological mechanism for coping with them".

The meaningful questions that need to be answered appear to be:
— at what levels do the pesticides occur in the various sectors of the environment;
— are these levels increasing or are they likely to increase;
— are these levels harmful, or are they likely to be harmful, to man or to other species.

1.2.1 Insecticide residues in atmosphere

Although occasionally measurable in parts per 10^{12}, organochlorine insecticides in the atmosphere are only of a minor or negligible importance (Breidenbach 1965, Tabor 1965, Abbott *et al.* 1965, 1966, Robinson *et al.* 1966, Stokinger 1969). Abbott for example reported a dieldrin level of 21 parts per 10^{12}.

1.2.2 Insecticide residues in water

Organochlorine insecticide levels in rain- and drinking-water are extremely small and under normal circumstances negligible (Wheatly *et al.* 1965, Abbott *et al.* 1966, Robinson *et al.* 1966). For instance dieldrin levels at 7 different locations in the U.K. were found to be 0-40 parts per 10^{12}. The dieldrin content of rain-water sampled in 1966/67 was lower than in previous years (Tarrant *et al.* 1968). According to these same authors aldrin and endrin cannot be detected in rain-water in the U.K. at detection levels of 1 part per 10^{12}. (This ratio on a time scale would be as small as to represent one second to more than 31,000 years).

Water of lakes and rivers could conceivably become contaminated from different sources:
a. leaching through soil;
b. transport of contaminated soil particles into waterways as a result of heavy rainfall or flooding;
c. fall-out of insecticides with rain-water.

From the literature (Wheatly *et al*. 1965, Abbott *et al*. 1966, Robinson *et al*. 1966, Tarrant *et al*. 1968, Stokinger 1969) it is apparent that a., b. and c. are insignificant and are not of toxicological importance to man.

d. direct contamination from farm and factory effluent. Surveys of water pollution suggest that it is not normal agricultural use, but effluent from industrial manufacture and formulation or from certain uses or misuses of insecticides (aircraft spraying, washing of utensils and accidental leakage from drums) that are the main sources of water pollution arising from insecticides.

1.2.3 Insecticide residues in soil

Data from the U.S.A. and U.K. show that on repeated application, the level of aldrin and dieldrin does not continue to build up in the soils treated. These data show that in temperate climates these insecticides after a single application disappear in an exponential way within four years; on repeated application of a given amount of these insecticides, year after year, the level builds up to a plateau (which is never higher than twice the level after the first application), or declines. This plateau is reached after 2 or 3 applications (Decker 1968, Lichtenstein 1968).

In the "Further review of certain persistent organochlorine pesticides used in Great Britain" (1969), the Report by the Advisory Committee on Pesticides and other toxic chemicals (the "Wilson-Committee"), it is stated in par. 277: "There is evidence that DDT residues persist and remain active in the soil longer than other organochlorine insecticides in common use, but it has been shown that, even where application of DDT or dieldrin have regularly been made to the soil, the residue levels measured have seldom exceeded the equivalent of a single application. Dieldrin, however, like any chemical or mechanical treatment, appears to affect some elements of soil micro-arthropod populations very markedly and for considerable periods of time. Nevertheless, most of the workers on this subject in Britain have considered that the resultant effects are not important to agriculture or the general environment".

After an initial application insecticides with a long persistence in soil are capable of producing residues in crops in successive years (Kraybill 1969).

According to West (1966) temperature, moisture, type of soil, amount of insecticide and volatility of the compound influence this persistence in the soil.

With the present level of usage of these insecticides there will be no further increase of their residues in the soil (Van Raalte 1965).

1.2.4 Insecticide residues in food and crops

Insecticide residues in food and crops are a direct result of the application of insecticides to crops growing in the field, and to a lesser extent from insecticide residues remaining in the soil. Residues in food of animal origin, especially in adipose tissues, may be, at a later date an indirect result of the same application. Because these residues form the main source of intake for man, residue tolerances have been established by governments and policing of residues in crops and food is being carried out.

Insecticide residues in food will be discussed in 1.3.1 and in the chapters on toxicology of the insecticides: it has been shown that insecticide residue levels in food in the U.S.A. do not increase (Duggan *et al*. 1966, 1967, Hoffman 1968) and in the U.K. residues are declining (McGill *et al*. 1969).

1.3   HEALTH HAZARDS TO MAN

1.3.1 Exposure of the general population

The different ways of exposure to chlorinated insecticides of the general — not occupationally exposed — population have been summarized by Campbell *et al.* (1965) and Kraybill (1969) in a scheme which is reproduced below in a somewhat modified form:

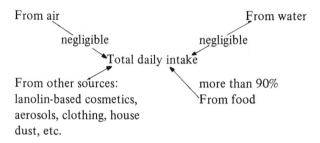

Air and water constitute negligible sources of organochlorine insecticides for man (1.2.1; 1.2.2; Stokinger 1969). For example, Campbell (1964) calculated that the DDT residues in the atmosphere and water combined contribute 0.04 mg per year to a total annual intake of 44.8 mg per man per year in the U.S.A. These calculation apply to normal U.S. and European exposures.

Food is the main source of exposure of the general population to organochlorine insecticides and accounts for more than 90% of total exposure (see chapter 5). Only in those areas where large amounts of pesticides are used for public health purposes (house spraying during campaigns against malaria, yellow fever, Chagas disease, filariasis a.o.) do these sources contribute significantly. Market basket surveys in the U.S.A. and the U.K. have shown that especially fat-containing foods such as fat, meat, butter, milk, etc. form the main dietary source of organochlorine insecticides. These insecticides are practically insoluble in water, but readily soluble in organic solvents and lipids. It is because of this latter property that the highest concentration of these insecticides is found in adipose tissue of animals and man.

1.3.2 Occupational exposure

Workers who are occupationally exposed to organochlorine insecticides during vector- or pest-control operations, agricultural usage, insecticide manufacturing and formulation, have an additional daily intake which may be 10-100 or more times the exposure of the general population (see chapter 13). This additional intake is mainly through dermal absorption and probably to a much lesser extent from inhalation of the insecticide (Wolfe *et al.* 1963). In this occupationally exposed group cases of intoxication are also more likely to occur than in the general population.

1.3.3 Health hazards

Stevenson (1966) sums up the possible risks arising from the use of pesticides as follows:

Acute risks, arising from handling and application of insecticides:

a. *Agricultural workers*: arising from the way in which the pesticide is applied.

b. *Factory workers*: intoxications arising in the course of the manufacturing and formulation process.

c. *General public*: in the home the accessability of poisons is an important factor; children are especially prone to intoxication from this cause. From time to time cases occur in which the general public is exposed to a toxic chemical because of the contamination of food during transit or due to a mistake in the identification of the product. These cases illustrate the necessity for adequate labelling and for transport and storage regulations.

Acute toxic effects in man have, in general, been the predictable consequences of accidental or careless gross overexposure (Zapp 1965).

Van Raalte (1965) collected from the world literature all cases (available at that time) of fatal poisoning by aldrin, dieldrin and endrin: 110 cases in all, and has tabulated them according to the mode of intoxication. The results are given in Table 1 below:

Table 1

FATAL CASES OF POISONING BY ALDRIN, DIELDRIN AND ENDRIN REPORTED IN THE LITERATURE UP TO 1965.

Sort of operation	Aldrin and dieldrin	Endrin
Spraying	1	4
Manufacturing	0	0
Accidental contamination	5	0
Accidental ingestion	3	24
Suicides	4	69
Total	13	97

From this table it is clear, that the great majority, almost 75%, of fatalities was due to suicide, and not to accidents. For comparison Van Raalte (1965) quotes the reported number of fatal parathion poisonings due to accidental contamination and ingestion over the same period: 3100 cases. The same author shows from available statistics on accidental poisoning in the U.S.A. and Holland that pesticides as a group and the organochlorine insecticides in particular are not an important cause of accidental poisoning.

Chronic risks

a. *General population*, which is exposed daily to pesticide residues in food.

b. *Workers* manufacturing or formulating insecticides or involved in large scale spraying operations from day to day.

From a toxicological point of view exposed workers are of considerable interest since medical observations on such people may provide important information for the detection of possible side effects from the use of pesticides (Stevenson 1966, Hayes 1966).

Effects from low-level long-term exposure have yet to be demonstrated in man if they exist at all (Zapp 1965). Before the health hazard from low-level long-term exposure of the general population to organochlorine insecticides can be assessed it will be necessary to establish:

a. the present body-burden of these insecticides in man;

b. whether this body-burden is likely to increase or decrease.

In chapter 5 it will be shown that the insecticide content of body tissues such as fat and blood is an indication of the body-burden of human beings, the previous intake level and the consequent hazard to health.

1.4 PUBLIC AND GOVERNMENTAL CONCERN

"Toxic substances have been used for offence and defence ever since man's advent on earth. Indeed such substances have played a large role in history, in romance as well as crime" (Stevenson 1966).

Concern about the effects of pesticides on health is nothing new. In 1919, pears with excessively high residues of arsenical compounds were seized by the Boston Mass. Health Department (Van Raalte 1964). Since then, particularly in the United States, the use of pesticides has been continuously under public debate and has frequently formed the subject for investigation and judgement by authorities.

Extensive scientific investigations into pesticide usage by the U.S. Government first started in 1925 and the Food and Drug Administration (FDA) held hearings on pesticide residues in the spring of 1930. From 1950 to 1953 exhaustive congressional hearings were held on chemicals in food. Throughout all the investigations two main subjects were discussed:

a. to establish the dose of a pesticide sufficient to control a pest;

b. to establish the amount of a pesticide residue in food which would not endanger public health.

Many countries now have organisations or committees for the establishment of residue tolerances, for policing pesticide legislation and for recommending changes in both to their Government. WHO/FAO activities in this field will be discussed later in this chapter.

Apart from concern expressed by Government authorities, public interest was raised by reports of side effects arising from accidents or the misuse of insecticides. This concern culminated in the publication of Rachel Carson's "Silent Spring" in 1962, which stimulated public interest in this whole subject. Her book promoted worldwide publicity and was followed by imitations.

The impact of Carson's book on the general public and Governmental bodies has been tremendous, by creating interest in problems concerning environmental pollution and pesticides and, indirectly, it has stimulated research in many fields. One direct effect has been a zealous demand to ban all pesticides, which of course would have been as unwise as unrestricted use of these materials. The pendulum has swung both ways and we are now in a period during which a new equilibrium may well be established.

It was not surprising that strong emotions pro and contra were expressed in the discussion especially when considering the fact that ornithologists were talking about toxicology, biochemists discussing ecology, etc., frequently without really understanding, or knowing the essential principles of the other discipline.

One of the more frequently quoted misconceptions remaining from that discussion is the misleading allegation that pesticide residues will continue to accumulate in human and animal tissues until death occurs. This is clearly not so. Nevertheless it is still widely reiterated by the uninitiated. In fact, it is a general principle of pharmacology and thus toxicology, that a steady state of storage will be reached as a result of continued, non-lethal intake of a drug or other chemical. At this steady state, a dynamic equilibrium is reached between the total intake of the insecticide per unit time, and the amount of pesticide metabolized and eliminated in the same unit of time (Hayes 1966, Kraybill 1969). The storage level reached under these conditions is dependant on the average daily intake (Hayes 1966, Hunter *et al.* 1967, 1969). Furthermore the levels of organochlorine insecticides found in the various tissues in the body are in direct relationship to the fat content of these organs (Fiserova-Bergerova *et al.* 1967) (See chapter 4 and 5).

After the "Spring" of 1962 and the subsequent publicity in the middle 60's, a report was prepared in the U.S.A. by President Kennedy's Science Advisory Committee on the possible hazards of insecticides and other materials that may or do pollute our environment. This report "Use of Pesticides" (1963) popularly called "Wiesner report" after the Committee's chairman, gave as its more important conclusions:

a. more research is required both on the effects of existing pesticides and the development of new and safer compounds;

b. ultimately, persistent toxic pesticides should be gradually withdrawn by orderly reduction in their use, when satisfactory alternative means of control have become available;

c. the Food and Drug Administration (FDA) should continue with its current review of residue tolerances.

From the more recent "Report of the Committee on persistent pesticides" (the "Jensen committee") (1968) in the U.S.A., the following conclusions and recommendations are relevant in this context:

1. Persistent pesticides are contributing to the health, food supply, and comfort of mankind, but, in the absence of adequate information on their behaviour in nature, prudence dictates that such long-lived chemicals should not be needlessly released into the biosphere.

2. Although persistent pesticides have been replaced in some uses and are replaceable in others, they are at present essential in certain situations.

3. Available evidence does not indicate that present levels of pesticide residues in man's food and environment produce an adverse effect on his health.

4. Persistent pesticides are of special concern when their residues possess – in addition to persistence – toxicity, mobility in the environment, and a tendency for storage in the biota.

The committee recommended amongst others that studies of the possible   long-term effects of low levels of persistent pesticides on man and other mammals be intensified.

Early in 1964, in the U.K., a report by the Advisory Committee on Poisonous Substances used in Agriculture and Foodstorage (Chairman Sir James Cook) to the Ministry of Agriculture, Fisheries and Food: "Review of the persistent organochlorine pesticides" was published. Primarily the questions to be answered were whether environmental contamination by these insecticides would have any undesirable effects on: 1. human beings and 2. wild-life.

The committee's answer to the first question was summarized in paragraph 130 of their report: "On grounds of human hazards, there is in our view insufficient evidence at present to justify a complete ban of any of the pesticides we have reviewed. There is, for instance, no basis for statements that these persistent organochlorine pesticides are severe liver poisons; nor is there any proof that DDT causes any injury while stored in the fat of human beings or animals. Similarly DDT and dieldrin cannot be condemned as presenting a carcinogenic hazard to man. The limited evidence available shows that, of the residues found in food, the maximum dieldrin levels were obtained only in isolated cases and under conditions which are seldom likely to arise. Nevertheless, we consider that these levels of dieldrin residues are undesirable and the evidence justifies a partial restriction of its use".

The committee's answer to the second question was summarized in paragraphs 132-134 of their report, which states, in short, that apart from circumstantial evidence for the view that the decline in population of certain predatory birds — particularly the Golden Eagle — is related to the use of aldrin, dieldrin and heptachlor and to some extent DDT, no evidence was received by the Committee that the populations of other species have been affected by pesticides.

These reports together form an up to that date summary of all aspects of the use of organochlorine insecticides.

Decker (1967) summarized it in one sentence: "In general there appears to be reasonable agreement that pesticides as used in agriculture and public health pose minimal, if any, hazards to man, his domestic animals, and his food supply, but that where wild-life is concerned, certain specialized pest eradication or suppression programmes do involve some hazards or calculated risks".

As to the general acceptability of this environmental contamination two Dutch experts may be cited here:

As a general principle, Zielhuis (1967, 1969-b) proposed for toxicants in general (freely translated quotation): "All toxicants have to be regarded in the first instance as foreign substances in the relation of man and his environment. As such, possible responses to toxicants in man may never be regarded as positive or desirable effects, at most they may be not-unacceptable. This does not imply, however, that every response as such is unacceptable. Unacceptable however are those responses which adversely influence human health".

"The fight against contamination of our environment will always remain full of compromises, since it is difficult to combine rapidly increasing prosperity with less contamination of the environment" (translated after Van Raalte 1969).

All the above-mentioned statements have recently been confirmed as a result of the careful evaluation (1969) by the Advisory Committee on Pesticides and other Toxic Chemicals in the U.K. ("The Wilson-committee"). Some of the paragraphs which are relevant in this context are quoted in full or in part, below:

89. "We consider it undesirable that the human environment should contain substances capable of producing toxic effects and whose continued presence conveys no benefits to human survival and well-being. If, however, priority is given to the removal of those substances the presence of which is known to be harmful to man, then on such a basis no high priority can presently be assigned to the removal of DDT or dieldrin. Nevertheless, we consider that evidence should be obtained by the regular determination

or organochlorine residues in people and their diets, so that the situation and trends can be kept under surveillance, and we recommend accordingly".

97....... "We believe that the problem of persistent organochlorine pesticides in the environment is only part of the larger problem of the general pollution of the environment by man; and although important, it should be looked at in terms of the priorities stated in paragraph 89 against the background of pollution arising from industrial and domestic effluents".

98. "Nevertheless, we are of the opinion that it is prudent to lower residue levels whenever possible".......

In paragraph 100 it is stated that for some situations the organochlorines are probably irreplaceable.

296. "In concluding our discussion on the outcome of this Review we wish to emphasize some very important points. Firstly the detection of these compounds in some sectors of the environment at levels as low as a few parts per million-million has been made possible by the very sensitive techniques now available. The use of these methods has shown that the more stable of these chemicals may accumulate in soils, though residue levels seldom exceed the level of a single application. Measurable amounts of dieldrin and DDT can be detected and confirmed in human body fat. There is therefore no doubt that the general population is being exposed to these chemicals but we have no evidence that this results in any adverse effects on man. Furthermore, the evidence we have about current exposure suggests that the levels in human body fat are not rising, and may be falling, and does not suggest that there are likely to be dangers in the long term, but clearly it would be wise to maintain a continuous check on the situation. We must remain in some doubt about the interpretation of the data on laboratory animals referred to in paragraph 86, the implications of which could be so important that we believe these data must be discussed, and agreement on their interpretation reached, on an international basis".

298. "We feel that, on balance, there is no case at present for the complete withdrawal of any of the chemicals under review. However, we shall continue to keep the situation under examination. As alternative pesticides are discovered and become commercially established, we shall consider the remaining uses of these organochlorine pesticides in the normal operation of the Safety Schemes".

1.5   THE ASSESSMENT OF HAZARDS TO HEALTH

As mentioned in 1.3.3 the assessment of the health hazard of the general population from low-level long-term exposure to organochlorine insecticides consists of two initial phases:
1. The establishing of the present body-burden of these insecticides in man, and determining whether this body-burden is increasing or decreasing.
2. Establishing whether this body-burden is harmful or likely to be harmful.

To this end we have to review successively:
a. Estimates of the current average daily intake of the general population.
b. The toxicology of the insecticide.
c. The evaluation of an acceptable daily intake, which may then be compared with actually measured exposures.

1.5.1 Measuring exposure of the general population

Until recently this exposure could be measured only by determining the insecticide content of the food consumed. This can be done in different ways, the most accurate of which are the market basket surveys and the studies of complete prepared meals.

Market basket surveys have been used for a number of years by the United States Food and Drug Administration (FDA). Samples of food are obtained regularly at retail stores in different geographical regions, prepared for consumption and analyzed. The insecticide content of the quantity of food calculated to be consumed by a 16-19 year old boy — this age-group consumes almost twice as much as the standard adult individual at moderate activity — is thus determined.

Determination of insecticide levels in complete prepared meals has been used since 1953 by the U.S. Public Health and subsequently in similar studies in Great Britain. In these studies complete prepared meals from prisons, hospitals, restaurants, etc. are analyzed and taken as representing the food intake of an average adult male at moderate activity.

These studies give an estimate of the actual intake per day in the general population, although it will be clear, that levels estimated in market basket surveys in the U.S.A. for the 19 year old boy will be much higher than those estimated for the average adult (See chapter 5).

For some years it has also been possible to determine actual levels of insecticides in body tissues. From a study of human volunteers, ingesting defined daily amounts of dieldrin, it has become possible to calculate the average daily exposure from a determination of the levels of insecticide in body tissues.

1.5.2 Toxicological study of an insecticide

Insecticides are by definition toxic products. However as with every product the degree of hazard depends on the dose ingested. A very toxic product may be handled safely (Hayes 1957); a relatively non-toxic product may cause harm if handled carelessly.

Toxicological studies should reveal the dose effect relationship on which the safe dose of a substance can be established.

Apart from animal-experimentation, the following studies involving man are of great importance in the assessment of the hazard to human health:
— a study of cases of accidental poisoning;
— human volunteer studies in which human volunteers are exposed to non-toxic doses which are, however, considerably higher than the intake of the general population;
— monitoring the health, clinical, biochemical and epidemiologic data of occupationally exposed workers. These workers are exposed for prolonged periods to dosages many times higher than those to which the general population is exposed. Thus this category should reveal the more subtle toxic effects of pesticides, if any, before they become apparent in the general population (Van Genderen 1964, Zapp 1965, Stevenson 1966, Hayes 1966, 1967, Kraybill 1969).

As Hayes (1967) wrote: "Even in volunteers or workers, it is frequently possible to search for possible effects of dosages from 100 to 1000 times those encountered by the general population. If such dosages produce no detectable injury in the volunteers or workers, the chance that some injury to the general population will ultimately occur is extremely small and in fact statistically negligible".

1.5.3 Establishment of an acceptable daily intake

Van Genderen (1964) described 3 phases in establishing acceptable daily exposures:

Phase 1: The toxicological study of the insecticide. Minimum requirements for this study are drawn up by the FDA in the U.S.A., the European Commission and by the Joint Meeting of the FAO and WHO — the latter giving advice for international use. Holland follows the requirements set by the European commission. This phase forms the scientific and toxicological basis for the following phases.

Phase 2: The establishment of an acceptable daily intake (A.D.I.) for man. The FAO/WHO (1968) defined the A.D.I.:

"The acceptable daily intake of a chemical is the daily intake which, during an entire lifetime, appears to be without appreciable risk on the basis of all the known facts at the time". It is expressed in milligrams of the chemical per kilogram bodyweight (mg/kg). This A.D.I. will generally be established at 1/100th of the level found to produce no effect in the most sensitive test animal, representing a possible tenfold difference in toxicity between members of the human species and another possible tenfold difference between man and experimental animals. Also taken into account is the fact that man is exposed to other toxicants in his environment. When sufficient data on human toxicity are available this safety factor may be lowered below 100. Theoretically, if full human data were available a safety factor of 10 — only accounting for a possible tenfold interindividual difference in toxicity in the human species — would suffice. A.D.I.'s are established by the joint FAO Working Party of Experts and the W.H.O. Expert Committee on pesticide residues (*e.g.* FAO/WHO 1967) and by national advisory committees in some countries.

Phase 3: The recommendation of residue tolerances for pesticides in food and crops. These are based on the average food consumption of the population and are set at such a level that the average daily intake will not exceed the A.D.I. However, establishment of tolerance levels is a matter of policy in which apart from toxicological data other factors play a role, as for example the state of nourishment of the population, the possibility of less hazardous alternatives, the level of environmental contamination, etc. Hatch (1968) stated: "...... a great deal of broad understanding and wisdom must enter into the fixing of tolerance levels quite beyond the limits of specific knowledge pertaining to the toxicity of the substances of concern". As Barnes (1967) wrote: "It is better that a population has some tissue fat in which the chemist can find high levels of chlorinated hydrocarbon insecticides, than that undernourished people die of diseases like tuberculosis but with tissue levels of DDT no greater than their wealthier neighbours".

Or to quote part of paragraph 299 of the Report of the Advisory Committee ("Wilson committee" 1969): "There is, of course, concern in other countries about the presence of pesticides in the environment; the extent to which account is taken of this elsewhere must clearly depend on the situations in countries which face problems in agriculture, public health, hygiene and living conditions, which may be quite different from our own. Thus, in tropical countries where food production is vital and there is a high incidence of mortality from insect-borne disease, the hazards to wild-life and the presence of minute residues in human fat may rightly be regarded as relatively unimportant. On the other hand, people in temperate countries with high living standards, and no threat from insect-borne disease, may reasonably be more concerned about the undesirable aspects of persistent pesticides".

1.5.4 Some comments on the actual situation

Barnes (1967) wrote: "Pesticide residues in food are not a health hazard and the constant watch kept on their levels and the acceptance of tolerances, legal or otherwise, serve more as a means of controlling the correct use of pesticides thus preventing outbreaks of poisoning among customers". The same author states that actually in the U.S.A. only 1 sample in 1000 food samples examined contains pesticide residues in excess of the legal tolerances. Most samples are far below the tolerance or contain no residues at all. These residue tolerances are the amount considered to be non-toxic even if they were consumed daily over a lifetime.

The FAO/WHO Joint Committee (1968) concluded that "at the present levels of pesticide residue intake the effects that pesticides can have on enzyme systems that metabolize other pesticides or drugs do not appear to present hazards to the consumer" and that "with the present levels of consumption as indicated by current total diet surveys, there is no need for concern about the possible additive or potentiative effects of an intake of more than one pesticide of the same group or of different groups".

At this point it is probably right to mention again the FAO/WHO definition of the A.D.I. (1968) cited in 1.5.3.

"The market-basket surveys show that food-contamination by pesticides is well below FDA tolerance levels" (Stokinger 1969).

1.6   PREVENTION OF RISKS FROM PESTICIDES

The problem of ensuring that pesticides are properly used is broadly one of education and training and as such it is not the responsibility of the manufacturers alone. It also requires the continuous efforts on the part of the authorities, expert advisors and the users.

West (1966) stated: "A society which expects to reap the benefits of its technological tools must learn to control the adverse side effects of these tools, whatever they may be. To control adverse side effects arising from pesticides better, the needs are most evident in research, field testing in monitoring the environment, in human health surveillance and in limiting the use of pesticides to persons who are competent by knowledge, training and equipment".

Barnes (1967) again stressed the need for proper training and proper use of equipment by applicators of pesticides and adds: "To ensure that the use of these materials presents no hazard, one obviously wants to apply the smallest dose which gives adequate pest control, which is economic to the farmer, and which leads to the least possible risk of food becoming contaminated".

1.6.1 Alternative methods of pest control

"The perfect pesticide would achieve complete control of the pest species, have an effect only on that species, quickly break down or otherwise lose its identity, and be economically acceptable" (Carman and DeBach 1967).
Steady progress is being made in the development of less hazardous insecticides.

Alternative methods of pest control at present available, include breeding of natural enemies of pest species, releasing of radioisotope-sterilized males of the species involved, the use of sex-attractants combined with chemosterilants, and many other methods. However these are unlikely to replace the conventional methods within the next decade from a practical point of view, because each of them must be tailor-made for each pest species, and their use must be possible from an economic aspect (Carman *et al.* 1967, De Boorder *et al.* 1969).

A survey of alternative methods of pest control is given in "Leven met insecten" (1969).

1.6.2 Voluntary preventive measures

Stevenson (1966) concluded: "The major sources of over-exposure relate to accidental ingestion or careless usage, factors which can only be overcome by extensive educational programmes in domestic, occupational and personal hygiene. The pesticide industry itself can effectively contribute towards this by ensuring that its products are adequately labelled, by recommending how unwanted stocks and empty containers can be rendered safe and by ensuring that their sales information contains easily understood recommendations as to safe usage. In particular it is necessary to emphasize the hazards of leaving chemicals within reach of children and animals and also the fact that twice the recommended dose not give twice the pest control".

An important requirement is the proper labelling of products. The label should contain all the necessary precautions for safe handling based on the knowledge acquired by the manufacturer during the extensive testing and development stage of the product. Safety instructions in the form of booklets and films should be made widely available, as well as instructions for use of safety equipment. Lectures and training courses for those handling insecticides need to be given.

An example of voluntary co-operation in prevention is the booklet "Pesticides, a code of conduct". This gives guidance to organizations and individuals who deal with, use, or study the effects of pesticides. It was issued as a joint effort by the U.K. manufacturers, distributors and users of pesticides, the Ministry of Agriculture, Fisheries and Food, the Natural Environment Research Council and Voluntary Conservation Organizations. A similar guide is under consideration in Holland.

The Shell-film "Read the Label' is one of the instruction films made available via the Shell organization or the FAO/WHO, to users of pesticides anywhere in the world.

1.6.3 Legal restrictions

"The object of the law is the safeguarding of men, both as users or applicators of a pesticide and as consumers of the treated crop. But the object is also the safeguarding of livestock, domestic animals and wild-life. This means that provisions in this legislation must be rather extensive, covering all phases of a pesticide from its manufacture through transport and storage, formulation, application and other handling, waste disposal, interval between application and harvest or between application and final consumption of the treated crop, and final tolerance on the raw commodity of the residue at the moment of consumption" (Van Raalte 1967).

"Contrary to what is sometimes alleged, industry appreciates that legislation is required. Generally speaking, industry welcomes the introduction of legislation and regulations and will be happy to co-operate with Governments. Such co-operation would be useful, since few people would know more about a particular pesticide than those who make it and who have lived with a compound for many years before it even comes into the market" (Van Raalte 1967).

It is however equally necessary to guard against over-reaction to incidents and against unreasonably stringent restrictions. The risk of using an insecticide must always be weighed against the benefits, the balance-level between the two will vary from country to country.

A working party from the FAO committee on Pesticides in Agriculture has drafted a "model-law", giving FAO's recommendations for the guidance of Governments wishing to introduce new legislation.

A survey of Governmental supervision and legal measures concerning pesticides in Holland was given by Flipse (1969), Flipse *et al.* (1969) and Van Genderen (1969).

1.7 SCOPE OF THIS STUDY

For the evaluation of the hazard of insecticides to man, data on human health are particularly needed since these are more relevant than the best animal experiments, following Alexander Pope's thought: "The proper study of mankind is man" (quoted by Coulston 1969).

"Some of the most valuable systematic studies on the relationship between exposure to noxious agents and consequent ill-health have been carried out in industry, making use of appropriate statistical and epidemiological procedures to relate levels and duration of environmental exposures with nature and magnitude of ill effects in exposed workers", and "As the problems of concern have become more subtle and complex, the need for expanding such industrial studies increases ......", both quotations from Hatch (1968).

To re-emphasise the point in 1.5.2: These occupationally exposed workers are exposed for prolonged periods to dosages many times higher — up to 100 or even 1000 times — than the daily dose to which the general population is exposed and thus this group of people should exhibit the more subtle toxic effects of pesticides, if any, in advance of the general population (Van Genderen 1964, Zapp 1965, Stevenson 1966, Hayes 1966, 1967, Kraybill 1969).

It is of course still impossible to collect human data from lifetime exposures to insecticides, but here it may be appropriate to quote what Hayes wrote in 1967, when considering medical data from DDT-workers: "I am not familiar with any drug or any other chemical, which has produced illness in the course of a lifetime when given at a dosage one twentieth of that which produced no effect when given for one-eighth of the life span of the species".

In the insecticide plant of Shell Nederland Chemie N.V. at Rotterdam-Pernis more than 800 workers have been handling aldrin and dieldrin for periods up to 15 years and endrin for up to 13 years. Medical data from these workers have been reported previously by Hoogendam, Versteeg and De Vlieger (1962, 1965).

The scope of this study is to present an epidemiological survey of all available data concerning this group of workers and compare their clinical data with their exposures in so far as it was possible to measure both. In our opinion the amount of data available is sufficient to enable conclusions to be drawn as to the acceptability of the exposure of the general population to these insecticides.

The parameters used in this study will be discussed first, followed by a survey of the toxicological literature on this group of insecticides.

The epidemiological study itself will be divided under three main headings:

a. The study of a long-term occupational exposure group up to January 1st 1968.

b. A study of all cases of insecticidal intoxication which have occurred in the insecticide plant.

c. A study of parameters of all insecticide workers up to January 1st 1970, in relation to insecticide levels in their blood, and as compared with a non-exposed control group.

1.8 SUMMARY

Different aspects of the pesticide problem have been reviewed briefly in this chapter, as a general background for the human health data which will be studied later.

In this chapter the scope of the study has been delineated.

Chapter 2

Parameters considered in this study

2.1   INTRODUCTION

2.1.1 Parameters for intoxication

The principle of medical supervision, including the necessary periodic examinations of workers who are specifically exposed to toxic hazards in their job is now well established.

The only objective must be the detection in an early pre-clinical stage of acceptable biological responses of the body to potentially dangerous materials. The detection of these responses in a pre-clinical, reversible and non-incapacitating stage allows the prevention of non-acceptable biological responses and progressive development of occupational disease with subjective manifestations, "complaints", and, finally, actual symptoms of disease. Actual disease is one of the last phases and a late result of a series of disturbances and compensations, which occur as an interaction between the body and a toxicant.

Early detection also helps to avoid hazardous exposure in other members of the group by allowing the elimination of further hazardous exposure. It is obvious that periodic medical examination cannot prevent an acute intoxication due to accidental gross over-exposure (Van Raalte 1966, De Bruin en Zielhuis 1967-I).

Thus, adequate preventive measures can be taken only if we know the *dose response relationship* which exists between, on the one hand, *the dose of the toxicant* which includes the exposure over a relatively short period as well as the pre-existing body-burden (see 2.1.3) and, on the other hand, *the effect of this dose on the subject.*

Of course, as Hatch (1968) poses: "To establish a basic scale of dose requires that the critical site of intoxication be known and that the mechanism of damaging action be understood in such fundamental terms as to distinguish clearly between the nature of the active agent operating at the critical site and the form of the external agent".

Measuring the dose of the toxicant will be described in 2.2 and 2.3. The effect of a certain dose on a subject depends on the inherent toxicity of the compound involved as well as on the individual susceptibility of the subject. In this context it must be remembered, that the toxicity of a compound, which is "the ability to cause harm", is something quite different from the hazard of this compound, which is "the probability to do so". This hazard depends to a large extent on the physico-chemical properties and on various environmental factors.

The measurement and/or estimation of the effects of a toxicant on the subject will be described in 2.4 and 2.5.

Every impending disturbance of the internal equilibrium of the body following on this interaction of toxicant on the body and body on the toxicant causes a compensatory reaction of the body.

As a result of exposure to a toxicant, chemical changes may be found in the body which may be schematically divided into:

a. The presence of levels of the toxicant in body fluids and/or body tissues — a selective parameter.

b. Presence of detoxification products of the toxicant which may be found in body fluids, tissues or in excretion products — selective parameters.

c. The occurrence of normally non-present biological compounds or signs as a response to the toxicant, such as biochemical change or, in the case of the insecticides under discussion, EEG changes typical of over-exposure. These again are selective parameters.

d. The occurrence of normally present biochemical compounds with altered levels or in different places than found under normal unexposed conditions. Examples are changes in serum enzyme activities and changes in the electrophoretic pattern of the serum proteins. These are non-selective parameters (De Bruin and Zielhuis 1967-I).

The distinction between selective and non-selective parameters, however, is not always as clear as suggested by the schematic enumeration given above.

Figure 1 shows the disability scale of a progressive intoxication with the effects plotted against time. This period of time may be expressed in minutes up to years depending on the toxicant involved (H_2S and HCN versus silica and heavy metals). It is not necessary, as regards running the full course of the disability scale, for the toxicant itself to accumulate continuously or significantly in the body, as long as the effects accumulate, as is the case with organophosphate insecticides and ethanol (Van Raalte 1966).

In *phase 1* there is a *normal adjustment* (Hatch 1962) to the toxicant with detoxication and elimination. Depending on the amount and the rate of absorption per unit time, the body can or cannot cope with the challenge to homeostasis. If it cannot,

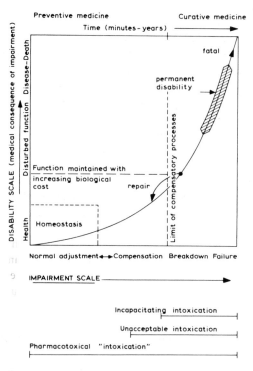

Fig. 1. The relationship between impairment and disability (modified after Hatch 1962–1968, and Williams and Zielhuis 1968).

some physiological functions will be disturbed. Which one depends on the target organ. This takes us gradually into *phase 2*: the *compensation phase* which is an increased effort for adjustment. Because organs have a considerable amount of reserve capacity, compensations for the impairment occur with normal body function being maintained, without undue effort and without the appearance of complaints or symptoms. However during this compensation phase, organ function tests, or enzyme determinations may indicate that the organism is somewhere in this compensation phase and may already be transgressing the state of homeostasis. Already from phase 1 on, as soon as there is a response of the body, there exists an *"intoxication" in the pharmacotoxical sense*. The mere presence of the toxicant or its metabolite, or the compensatory reaction of the body, however, does not necessarily mean that this is harmful, and although it may not be desirable, it may still be not unacceptable. The detection of a toxicant in body tissue definitely is not identical with unacceptable intoxication. The acceptability depends on the quality and quantity of the response measured. During this phase there need not be any manifest signs or symptoms, however, during the latter part of this phase some atypical complaints may occur. A non-specific response may arise, common to all kinds of physical and chemical stresses, such as fatigue, headache, loss of appetite. As soon as this effect has reached the point where it is no longer regarded as acceptable it must be regarded as an unacceptable intoxication which, however, is not necessarily identical with incapacity, but only means that something has to be done to prevent the occurrence of the *last phase* of *the incapacitating intoxication* (Hatch 1962, Van Raalte 1966, De Bruin and Zielhuis 1967-I, Hatch 1968, Williams and Zielhuis 1968, Ariens 1969-b).

Now by experimentation, observation and experience we have learned for a few chemicals which concentration in the body is associated with a certain degree of exposure on the one hand, and with what phase of intoxication on the other. In the same way we have established the relation between selective and non-selective parameters. In many cases we still have to rely on the use of non-selective parameters only, which implies that conclusions can only be drawn from the examination of groups of workers.

From the foregoing and from Fig. 1 it will be clear, that selective examination with selective parameters is of utmost importance in preventive medicine, not only from a point of view of selectivity, but also as it makes much of routine physical examination, blood counts and urinalysis and a lot of biochemical and enzyme determinations superfluous in routine examinations of the workers exposed. This of course is possible only when the above mentioned relationship between parameters selective for the toxicant and non-selective parameters has been well established, which also means that the points of transition between the phases of normal adjustment, compensation and breakdown are known. On farther sight this threshold for industrially exposed workers allows an adjudication — together with other data — of the acceptability of the exposure of the general population.

According to the objective of preventive medicine we are, of course, aiming at keeping the health of our workers in the lower left-hand rectangle (of Fig. 1), that of normal adjustment and health. What can be achieved is illustrated by our insecticide workers (see chapter 11): we started off with convulsive incidents and clinical intoxications, followed by a period of preventive routine EEG testing with secondary prevention. Nowadays using the determination of insecticide levels in the blood and keeping them below established threshold levels we have passed on to the stage of effective early prevention. Of course this method currently in use cannot prevent acute intoxications from accidental gross over-exposure. In the prevention of accumulative intoxications, however, it proved to be extremely successful.

2.1.2 Individual factors involved

Apart from the dose and the inherent toxicity of a drug or a chemical, the individual susceptibility of the subject plays an important role in determining the effect of this dose on this individual. This individual susceptibility is very complex (Zielhuis 1960). The factors affecting toxicity include:

Age

It is known for instance that some detoxication mechanisms in newborn individuals are not as well developed as in adults (Conney 1969). With DDT and dieldrin this may not be the case. Microsomal enzymes involved in DDT metabolism are probably fully functional at birth, although no direct studies are available to confirm this. However, in the mouse (Catz *et al.* 1962) and rabbit (Hart *et al.* 1962) preparturition dosing with phenobarbitone (Wright and Donninger 1968) has resulted in marked stimulation of the microsomal enzymes. In acute studies in the rat (Lu *et al.* 1965) dieldrin has been shown to be less toxic in the neonate than in the adult.

Sex

There may be sex related differences in detoxication such as occur in rats and other test animals with some compounds.

Heredity

Hereditary defects in metabolic pathways may exist ("inborn errors of metabolism"). Allergic diathesis may influence individual reaction patterns (skin, respiratory tract, blood, etc.).

General health

Pre-existent or former disease are self-evident factors in this respect. The condition of the liver as the main biotransformation centre of the body is of utmost importance in workers in the chemical industry. Thus chronic alcoholics are notoriously more susceptible than others: for this reason and perhaps also on account of their mental state (*post aut propter*). The condition of the effector organ of a drug or chemical is also very important.

Nutritional state and constitutional type play a role. The amount of body fat is important in persistent lipid soluble chemicals, as this is the main site of storage (in a biological inactive state for the insecticides under discussion). Body weight, body length and the relation of these parameters are of importance here as indicators.

Simultaneous exposure at work or at home to other chemicals and/or drugs may be of influence, for instance through potentiation, addition, antagonism, enzyme induction. There is also the possibility that residues from previous exposures are still present or that the body has not yet fully reversed the effects of such previous exposures. In the same way environmental stresses (cold, hot climate, noise) may deter the biological response.

2.1.3 Accumulation, body-burden, biological half-life

The blood usually serves as a medium for distribution of a drug or a chemical through the body irrespective of the porte d'entree. The blood will take it to the organ where biotransformation takes place, to the effector tissues or organs where a specific response can occur, to those tissues which may store the drug or chemical, and to those organs which will excrete it.

As a rule elimination of a drug or chemical which includes detoxication and excretion is exponential, which means that per unit time a certain percentage of the then available amount is excreted which is typical for that drug or chemical. The *biological half-life* (t½) of a compound is the time needed to eliminate 50% of this compound from the body. The same time is then needed to eliminate half of the remaining 50%, and so on.

Hydrophilic compounds usually have a short biological half-life. Lipophilic non-polar compounds have a longer half-life, mostly caused by their inactive storage in body fat.

When a chemical has a long half-life and/or its daily intake is high (*e.g.* industrial exposures) the possibility arises, that during certain periods the intake exceeds elimination. Then *accumulation* of the chemical in the body occurs.

When accumulation occurs, the drug or chemical is stored in the body. For lipophilic compounds this storage will take place in body fat and if the levels of these compounds in different tissues or organs are compared, one finds that there is some correlation between the storage in a certain tissue and its fat content (Fiserova *et al.* 1967), with the exception of the liver where relatively higher concentrations are found, probably due to its function in metabolism and excretion of these compounds (for organochlorine insecticides). This distribution was confirmed for dieldrin by De Vlieger *et al.* (1968): see chapter 5.4.2.

When accumulation has occurred the *total body-burden* means the total amount of a drug or chemical in the body at a certain time. Consequently, the body-burden is the result of the intake by all routes of entry, the metabolism and the elimination as characterized by the half-life of the compound (Zielhuis 1969-b). This whole chain of events occurring when a foreign substance enters the body — the pharmacokinetics of the substance — and not included the response of the body, is shown schematically in Fig. 2 (Williams and Zielhuis 1968).

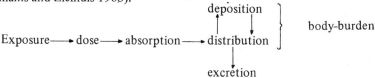

Fig. 2. Schematic chain of pharmacokinetics in the body (Williams and Zielhuis 1968)

As elimination, according to the general principle described above, increases with the amount of chemical or drug available in the body, a point will be reached where intake and elimination are equal again (for that level of daily intake). At this stage of equilibrium between intake and elimination a steady state of storage (Hayes 1967) is reached with a constant total body-burden of the toxicant — the accumulation level — in dynamic equilibrium with a corresponding constant blood level for the toxicant and constant levels in other tissues of the body. When the mathematical formula for these relationships is known (as it was determined by Hunter *et al.* (1969) for dieldrin), then at this state of equilibrium, determination of the level of the toxicant in blood will make it possible to estimate the total body-burden as well as the average daily intake of that toxicant (see chapter 5).

So accumulation and elimination both take place exponentially and both are related to the half-life of the chemical involved. Reaching of the steady state and therefore also total elimination of a compound takes longer when the daily intakes are higher. This is illustrated in Fig. 3.

Whether or not a certain daily intake will have a toxic effect in the long run depends on the margin between the steady state level reached and the level of the compound required to cause toxic effects. Thus a body-burden of many times the acute toxic dose of a certain compound may be without toxicological hazard, if the corresponding blood level — the biologically active material in the body — is safe.

Every chemical has its own characteristic biological half-life which, however, should be regarded as an average value. Interindividual variations may be considerable, for this half-life is influenced by many factors (Conney 1969, Ariens 1969-b), such as:
— Heriditary constitutional variations.
— The integrity of the detoxication mechanisms which is of the utmost importance and

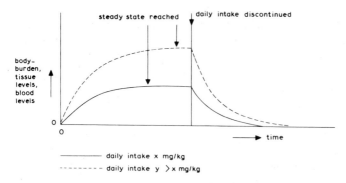

Fig. 3. Accumulation, steady state and elimination.

may be disturbed if liver function is failing (increased half-life).

— Interference with other drugs or chemicals at binding sites to plasma proteins or in detoxication systems (decreased or increased half-life).

— Enzyme induction caused by other drugs, chemicals, stresses or the compound itself — to be discussed in 2.1.5 — which may decrease the half-life considerably.

A striking example of marked decrease by enzyme induction is the following: the half-life of zoxazolamine is 9 hours in control rats, 48 minutes after pre-treatment with phenobarbital, and 10 minutes after pre-treatment with 3,4-benzpyrene (Conney 1967).

Enzyme induction not only alters the duration, but also the intensity of drug action (Conney 1969). Chronic administration of a drug may not only stimulate the metabolism of other drugs, but also, in many cases may lead to enhancement of its own metabolism, which explains the phenomenon of tolerance (Conney 1969).

— Reduction of microsomal enzyme activity, on the other hand, increases the half-life. Examples of inhibitors of microsomal enzyme activity are: carbon tetrachloride, 3-aminotriazole, nicotinamide and such normal body substrates as the pituitary hormones somatotropin, corticotropin and prolactin (Microsomes, etc. 1969).

As the biological half-life of a drug or chemical determines the level of the steady state of accumulation which will be attained and this again finds its expression in the duration and intensity of a drug action or an intoxication, it is clear from the foregoing that there might be a comparatively wide interindividual variation in the reaction to drugs or chemicals.

2.1.4 No-effect level

After having discussed in the foregoing sub-chapters individual factors in the reaction of the subject to a toxicant, the principles of the influence of the toxicant on the subject will now be discussed.

Generally for every drug or chemical there is a direct quantitative relationship between the dose of intake and the response of the subject, but the shape of the curve representing this relationship depends on the body function, the symptom or the parameter measured (Zielhuis 1969). Fig. 4 illustrates some possibilities of dose response curves. This general dose response relationship also applies to pesticides (Hayes 1967).

Another factor which depends on the parameter used is the *threshold dose* for this effect: the dose at which this effect can first be measured. Apart from individual factors,

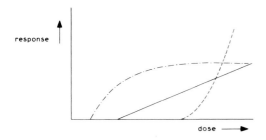

Fig. 4.   Different forms of the dose response curve.

most important here is the sensitivity of the parameter and the technique used.

All points where dose response curves cross the x-co-ordinate in Fig. 3 are *no-effect levels* for a specific parameter measured. This no-effect level tells us nothing about the meaning or its implication for the health of the subject. Certainly, with improved and more selective laboratory techniques it will be possible to detect effects at lower doses. Stevenson (1966) writes: "One of the areas of contention in experimental toxicology is the definition of the term "no-effect", for there are philosophical grounds for arguing that any exposure to any compound will produce an effect, if you know what to measure" and "Many of the criteria which are used in experimental procedures are in fact measuring exposure rather than an inherent toxic effect. As a result of this, the "no-effect" intake which is used to calculate an acceptable daily intake for man, may be extremely conservative".

On the other hand – and now returning to 2.1.1 and Fig. 1 – we do not want to reach the *toxic-effect level*, meaning the dose at which the phase of the non-incapacitating unacceptable response will be reached.

Here is a problem caused by improved laboratory techniques which stimulated the introduction some years ago of the concept of the *no-adverse-effect level*, which means the dose at which by all means and methods no adverse effect can be detected. Unfortunately this shifted the consideration of the actual problem to a discussion of the difference between adverse- and no-adverse-effect. Some authors state that every measurable effect is an adverse effect. Rather than entering into this unprofitable discussion it would be better to cite Roe (1968) "The problem of knowing how to distinguish between a toxic effect and a non-toxic effect is not only unresolved, but probably unresolvable". Hatch (1968) and Oser (1969) arrive at a similar conclusion.

Which of these levels will be used for administrative decisions, however, is clearly a matter of choice.

In our opinion, however, the no-adverse-effect level is a sensible level to consider and use for establishing acceptable intakes, the more so as a large safety factor is applied in arriving at an acceptable intake for men from a no-adverse-effect level in the most sensitive and appropriate animal species tested. We might add at this point that when we are able to determine a no-adverse-effect level in man a different but smaller safety factor could be applied.

The sometimes comparatively wide interindividual variation in reaction to drugs and chemicals has been mentioned above. The no-effect level, however, is based on data from groups, thus on the most sensitive members of these groups. Thus sufficiently large groups are needed to obtain statistically significant results.

2.1.5 *Enzyme induction*

As regards enzyme induction it would suffice to refer to the summary given by Conney (1967) who in his review article on this subject covered 379 publications. He writes in his summary:

"In increasingly large numbers, drugs, pesticides, herbicides, food additives and environmental carcinogenic hydrocarbons are being found to stimulate their own metabolism or the metabolism of other compounds. The evidence suggests that foreign chemicals exert this action by increasing the amount of drug metabolizing enzymes in liver microsomes. Treatment of animals or man with suitable inducers of liver microsomal enzymes accelerates drug metabolism *in vivo* and alters the duration and intensity of drug action. For instance, barbiturates decrease the anticoagulant activity of coumarin anticoagulants by accelerating their metabolism. This effect requires that the dosage of the coumarins be raised to obtain an adequate anticoagulant response and serious toxicity can result after combined therapy with a coumarin anticoagulant and a stimulator of drug metabolism when the enzyme stimulator is withdrawn and the anticoagulant is continued without an appropriate decrease in dose.

The stimulatory effect of drugs on their own metabolism often allows the organism to detoxify drugs more rapidly. This effect has considerable importance when it causes drugs to become less toxic and less effective during prolonged administration. However, if a metabolite has more activity than the parent drug, enzyme induction can enhance the drug's action. Enzyme induction may also be important during chronic exposure to environmental carcinogens, such as 3,4-benzpyrene. The ability of 3,4-benzpyrene to stimulate its own metabolism in liver, lung, gastrointestinal tract and skin represents an important mechanism for the detoxification of this substance.

Inducers of microsomal enzymes stimulate the metabolism or synthesis of several normal body substrates such as steroid hormones, pyridine nucleotides, cytochromes, and bilirubin. Evidence has accumulated that steroids are normal body substrates of drug metabolizing enzymes in liver microsomes. Accordingly, treatment of rats with phenobarbital enhances the hydroxylation of androgens, estrogens, glucocorticoids and progestational steroids by liver microsomes. This effect is paralleled *in vivo* by enhanced metabolism of steroids to polar metabolites and by a decreased action of steroids such as estradiol, estrone and progesterone.

Recent studies suggest that inducers of liver microsomal enzymes enhance the hydroxylation of steroids in man. Phenobarbital, diphenylhydantoin and phenylbutazone are examples of drugs that stimulate cortisol hydroxylase activity in guinea pig liver microsomes and enhance the urinary excretion of 6β-hydroxycortisol in man. Further research is needed to learn whether the stimulatory action of drugs on the metabolism of normal body constituents is harmful or whether it restores a homeostasis that was upset by drug administration. It is of considerable interest that certain inducers of liver microsomal enzymes have recently been used therapeutically for the treatment of hyperbilirubinemia in jaundiced children and for the treatment of Cushing's syndrome.

Considerable further work is required to evaluate more completely the effects of liver microsomal enzyme inducers on the metabolism of bilirubin, cortisol, and other normal body constituents in experimental animals and man".

Kupfer (1967) concludes that the administration of organochlorine insecticides to animals of various species stimulates both the hepatic microsomal oxidation of drugs and the microsomal hydroxylation of steroids. Conversely organophosphorus insecticides inhibit these reactions. This conclusion however is based on data relating mainly to DDD, DDT, chlordane and perthane. Aldrin, dieldrin and endrin are barely mentioned. Nowhere are the dosages quoted against the responses registered.

Although in many responses combinations of organochlorine insecticides appear to be additive, there are indications (Street *et al.* 1969), that structurally related substances like some of the chlorinated cyclodiene epoxides, when given in combination at near threshold doses (the threshold for this effect), might be competitive for microsomal enzyme systems, meaning that they retard each others removal from the body, as contrary to otherwise occurring enhanced detoxification. This however depends on the combination of pesticides and the doses used and there seems to be a wide variation in species sensitivity to these effects.

Zielhuis (1969-b) suggests that, although probably not relevant for the general population, enzyme induction might play a role in workers occupationally exposed to organochlorine pesticides.

Thus, enzyme induction is a reaction to many — mainly lipid soluble — drugs and chemicals. The induction is not related to the chemical structure of the compound, but the chemical fit between the compound and the biologic stucture upon which it acts, plays an important role: this determines whether hydroxylation, deamination, dealkylation, sulfoxidation etc., or glucuronide formation will be stimulated.

There is a dose effect relationship in the phenomenon of enzyme induction and without doubt there is a threshold level below which the phenomenon does not occur. These dietary thresholds have been established for some compounds including DDT and dieldrin, usually using the rat as the experimental animal. Thresholds, however, vary according to the species used, and: "Our limited work, chiefly oriented, on the DDT-dieldrin interaction, indicates that the rat may be a poor representative species because of its exceptional sensitivity" (Street *et al.* 1969).

Depending on the degree and the duration of exposure, electronmicroscopic examination may first show a proliferation of the smooth surfaced membranes of the endoplasmatic reticulum (SER) of the liver cells, and not of the rough surfaced endoplasmatic reticulum which contains the ribosomes. In association herewith there is an increase in microsomal protein synthesis and an increase in microsomal enzyme activity. This smooth surfaced endoplasmatic reticulum of the liver cells contains the enzyme systems for detoxification and probably also the excretion system for the hydrophilic compounds derived as a result of the detoxification process.

With increased dose or duration of exposure all structural elements of the liver cells increase to the same extent, only the smooth endoplasmatic reticulum shows the above mentioned extra increase. Sometimes even the rate of liver cell mitosis may be increased. All this results in an hypertrophy of the liver which is regarded as a sign of adaptation and is different from toxic liver injury or fatty degeneration. All the above mentioned responses of the liver are reversible phenomena, which return to normal after discontinuation of exposure (Hart *et al.* 1965, Stemmer *et al.* 1966, Gilbert *et al.* 1967, Remmer 1967, Conney 1967, Ariens 1969-a and 1969-b, Conney 1969).

"Almost every aspect of enzyme induction is potentially important to human welfare". On the one hand there is a possibility of undue interference with drugs and normal body substrates, on the other, the possibility that a certain degree of enzyme induction will counteract and protect man against environmental carcinogens and other toxicants and accelerate depletion of foreign substances such as insecticides from the body (Conney 1967).

Popper (1967) raised the question whether a structural deviation from the normal necessarily represents an adverse effect. He suggested using henceforth the term "no-adverse-effect level" rather than the "no-effect level", hitherto accepted by toxicologists.

Essentially the same opinion was expressed by Durham (1967) when he stated "The histologic effects produced in the livers of rats fed low levels of DDT or other organochlorine pesticides are another evidence of adaptation".... and "....when viewed as an adaptive change rather than a toxic effect, the histological effects on the rat liver of long-term ingestion of small amounts of organochlorine pesticides could then be considered as evidence not of harm from exposure to low levels of these agents, but of increased capability of the organism to cope with its environment". The FAO/WHO Working Committee (1967) expressed a similar opinion (WHO Techn. Rep. Series 348) when they wrote: "It is reasonable to believe that this alteration (*i.e.* enlargement due to chemical compounds) may not always represent a pathological change and in some instances, on investigation, be revealed to be a normal response to an incurred work load".

The FAO/WHO Working Committee (1968) in "Pesticide residues" considering the possibility of enhanced drug metabolism by organochlorine pesticides states: "This could be considered a potentially adverse effect of this class of insecticides. However, in the relatively steady state that exists with respect to the levels of these substances in the environment sudden changes in drug metabolism and interference with schedules of drug therapy would not be anticipated".

The FAO/WHO Joint Meeting thereby elaborated upon their earlier (1967) opinion: "Considering interaction effects between various pesticides and pesticides and drugs, it was agreed that at present levels of consumption of pesticides no alteration in A.D.I.'s or tolerances are required to take into account the possible additive or potentiation effects".

The whole mechanism of enzyme induction has not yet been completely investigated, and particularly the role which organochlorine insecticides might play in it in man is still unsettled.

This brings us to the theory of "Sufficient challenge" (Smyth 1967) which poses the inherent need of every body function to be sufficiently stimulated in order not to atrophy (Baetjer 1964). Examples of apparent benefit from small doses are common in animal experiments, but are usually attributed to artifacts. In many chronic studies on pesticides, animals receiving small intakes have been healthier than control animals (Stevenson 1966, Smyth 1967). Thus, many signs point to the fact that a certain degree of enzyme induction might be beneficial for general health, or, in other words, that between the "no-effect level" and the "no-adverse-effect level" there might be a level of intake at which a beneficial effect as a response to a foreign substance or stimulus predominates. In the case of environmental contaminants, such as some of the organochlorine insecticides, there are no indications in the world literature that threshold levels for any of these responses have been reached or are likely to be reached, as far as the general population is concerned. In fact, a simple calculation by Street *et al.* (1969) shows that actual dietary levels of organochlorine insecticides in the general population are far below the threshold levels for hepatic enzyme induction in the rat which is exceptionally sensitive for this effect.

However, there still remains the question whether such a "sufficient challenge" should be offered through forced intake, *e.g.* food additives, whereas normal life probably already offers many chemical stimuli. The question is unsettled whether or not the capacity for adaptive reactions is limited and should be spared as much as possible.

2.2   MEASURING EXPOSURE

2.2.1 Routes of exposure

Measuring exact exposures to individual insecticides was virtually impossible until quite recently. It was even more difficult to determine exposures in workers handling various insecticides at the same time or at different times.

The different routes of entry into the body have to be taken into account:
— Skin: the foremost porte d'entree in occupationally exposed workers in insecticides.
— Respiratory intake may play a role depending on circumstances.
— Gastrointestinal (oral) intake normally is of small importance in industry.

In every instance, but especially in the case of insecticides three modes of exposure exist:
1. Occupational exposure: industrial (manufacture and formulation), agricultural use, public health use and pest extermination.
2. Exposure when using the same material avocationally, such as in gardening.
3. Exposure through environmental contamination by their presence in food, water and air.

2.2.2 Factors influencing these routes of exposure

Intake through the skin is influenced by:
— the condition of the skin: dermatitis and eczema decrease the effect of the skin barrier and increase the uptake;
— increased circulation and sweating enhance the intake at a given exposure;
— skin hygiene: washing of contaminated skin and changing of contaminated clothes may be major factors.

Respiratory intake is influenced by many factors (Zielhuis 1969):
— the respiratory volume per minute determines the amount of toxicant taken in per unit time. This is increased:
— with physical effort, relatively more in smaller or untrained people;
— with disturbed cardio-respiratory function;

— in children having a higher alveolar ventilation per kg than adults.

— the degree of air pollution and the way in which the toxicant is present in the air: vapour, dust, mist or particles. This again is dependent on:

— the microclimate a worker creates around himself as part of the variations which exist in the total work environment.

Gastrointestinal (oral) intake may be influenced by:

— contamination of food by accident or poor hygiene;

— smoking during work, etc.

Apart from these variables, exposure depends mainly on two factors:

— duration of exposure and

— degree of exposure (concentration in air and skin contamination).

2.2.3 Duration of exposure

The duration of exposure when measured as the total time worked in a certain plant does not take into account how long a man actually is in contact with one or more of the insecticides and how long he is working at places where the exposure is much lower or non-existent. A more accurate estimation of duration of exposure is obtained by maintaining detailed work records of all workers concerned (Hayes and Curley 1968). These work records list for each worker the number of hours spent on the different jobs, specified according to the insecticides handled. Hours spent on sick-leave, holidays, etc. are also recorded.

From this it will be clear that the duration of the exposure is at best a very crude parameter.

It is obvious that extra hours worked overtime increase the total exposure and decrease the time for detoxication (Zielhuis 1960).

2.2.4 Degree of exposure

The degree of exposure is a still more complex factor to determine. It depends on:

Individual factors:

— the way in which a man conforms to the rules of personal and plant hygiene;

— the way in which a man creates his own microclimate: carelessness at work; creating more toxic dust than necessary etc.;

— new and inexperienced workers may create a microclimate not only hazardous for themselves, but also for those working next to them;

— the production level and the degree of individual physical effort required, influencing the alveolar ventilation.

Technical factors:

— the physico-chemical characteristics of a compound: vapour, dust, mist or particles, etc.;

— other products used as vehicle *e.g.* solvents, may facilitate penetration of a compound through the skin;

— the production process with its seasonal fluctuations in production rate and occasional peak exposures due to disturbances in the production process or the technical installation;

— the location and the kind of work: exposures are highest in our formulation plant,

somewhat lower in the production plants and still lower in intermediate production. Many workers move from one to another of these plants more or less frequently.
Inside the plants there are also differences: *e.g.* the man at the centrifuge of the technical product has higher exposures than the panel operator. Maintenance workers and clean-out workers mostly have exposure peaks only;
— the work climate: the temperature at work and the ventilation of the plant. The fact that manufacture is a continuous process (shift-work) influences the work climate: toxic dusts, etc. (insofar as they have not been removed by exhaust ventilation) do not have the opportunity to settle; cleaning of the plant must be done while it is still operational (Zielhuis 1960).

Methods used in determining the degree of exposure:
— Routine air sampling with static samplers. This has not been done in our insecticide plant. However, some 300 air samples have been taken at different times and different plant locations. Some of these results were mentioned and tabulated by Hoogendam *et al.* (1965). Comparison of data of 1960 with those of 1958 shows a sharp decrease of aldrin and dieldrin levels as measured in air samples of the insecticide plant. Even when done as a routine method, we would regard it as a crude indicator of overall exposure rather than as an indicator for individual exposures, which depend far more on changes in location and type of work and the individual's effort and attitude towards the work, as was mentioned above. Neither of course do air concentrations allow an estimate of skin contamination.
— Personal monitoring, in which the worker wears the air sampling module on his body, and which at all times and places registers the personal microclimate of the worker, should be mentioned as an important improvement on the foregoing method. We have not used personal monitoring in our insecticide plant because prior to this method becoming available to us we could use the blood control described below.
— Dermal exposure can be estimated by attaching absorbant cellulose pads to various parts of the body for one or more complete work cycles and determining the toxicant level in these pads; in the same way respiratory exposure can be estimated from the contamination of filter pads of respirators worn during work with a toxicant (Wolfe *et al.* 1963). Neither method has been used in our plants.
— Determination of the levels of the insecticides in the blood of the workers. Contrary to the methods described above, this biological and selective method is an exact indicator of the total exposure by all routes of entry: skin and pulmonary as well as oral exposure. This method was introduced in our insecticide plant in 1964 and has been used as a routine method ever since (see chapter 13).

2.2.5 Summary
It will be clear from the above discussion that we passed from a period in which air-monitoring was done occasionally by our chemical laboratory in co-operation with an industrial medical department who were looking out for the first signs of intoxication, into a period of biological monitoring with a close contact between chemical, biochemical and clinical laboratory. This enabled the industrial medical department to consider these data together with data from the general physical examination and to formulate a more rational advice with the objective of early prevention.

This biological monitoring has the great advantage over all the other methods mentioned in that the result is not dependent on the way of entry into the body or irregularities in place and time of exposure (De Bruin and Zielhuis 1967-I). Total exposure of the subject is measured and the method and frequency of monitoring can be adapted, among other things, to the half-life of the compound under consideration.

2.3   SELECTIVE PARAMETERS

2.3.1 Electroencephalography

As the central nervous system is the effector organ for this group of insecticides, changes in the electroencephalogram (EEG) may occur upon absorption of high doses of these insecticides.

Thus electroencephalography is useful for indicating over-exposure to these insecticides. However, the changes noted are those of stimulation of the brainstem and are not specific for over-exposure to these insecticides; they may be caused by any substance which stimulates the brain in this way. In fact EEG changes alone never allow the diagnosis of any specific disorder. An EEG may show changes in frequency, amplitude and phase-relations, which can be interpreted in association with an anamnesis and data of the physical examination of the patient (Hootsmans 1962).

In our insecticide workers a pre-exposure EEG is taken, so that subsequent EEG's can be compared against these "base-line" tracings. Changes in the pattern on later occasions are regarded as of more importance than the pattern itself. After insecticide intoxications, return to the "base-line" of the EEG is a good indication of recovery.

These changes involve bilateral synchronous thêta-wave activity and occasional bilateral synchronous spike and wave complexes thought to be associated with alterations in the function of the brainstem. Similar EEG changes may be found in non-exposed people. The difference is that some controls show this pattern permanently, whereas in insecticide contacts the pre-exposure pattern is normal and the pattern will return to normal again when the insecticide contact is discontinued (Hoogendam *et al*, 1962, 1965).

The appearance of EEG changes in exposed subjects with a previously normal EEG, however is a serious symptom of intoxication and must be regarded as a warning of imminent convulsions (see chapter 16).

From 1962 onwards, workers who showed EEG changes on their pre-exposure EEG were not admitted to work in the insecticide plant in order to avoid all possible future arguments of cause and effect relationship.

2.3.2 Insecticide levels in the blood

A very great improvement in the medical supervision of insecticide workers and the prevention of intoxication occurred when it became possible to determine the insecticide concentrations in the blood by gas liquid chromatography (GLC) with electron capture detection, as developed and described by Goodwin *et al.* (1961), Robinson (1963), Richardson *et al.* (1967) and Robinson (1969).

Measurement of the blood level of insecticides at the state of equilibrium reflects total absorption of insecticides by all routes of entry as well as the storage level in the adipose tissue. Thus it is a good measure of the total body-burden of insecticides. It appears to be far more practical for routine use than determination of insecticide levels in adipose tissue.

Such relationship has been established and confirmed for pp'DDE as the stable metabolite in the body of the more transient DDT (Davies *et al.* 1968, 1969-b, Edmundson *et al.* 1969), and for HEOD (dieldrin) resulting from the combined intake of dieldrin and aldrin, which latter compound is rapidly epoxidized into dieldrin in the body (Hunter *et al.* 1967, 1969) (see chapter 5).

For endrin (Korte 1966, 1967, 1968) and Telodrin (Korte 1963, 1967, Worden 1968) similar relationships were found in experimental animals, and these are not in contradiction with our own experience with workers exposed to these insecticides.

This combination of GLC with electron capture detection is a very sensitive analytical method for the determination of chlorinated hydrocarbon insecticides, which can now be detected in minute concentrations with a precision of approximately 5%, at levels of 0.001 μg/ml (= 1 ppb= 1 part per billion = 1 part per 10^9). On a time scale 1 ppb is equivalent to one-third of a second in 10 years, or expressed on a distance scale it is less than one centimeter in the circumference of the earth. It is important to realise this, because so often our ability to measure these very small quantities is confused with their significance and our ability to interpret them.

The detection levels in our laboratory are not quite as low as described above, but still sufficiently low from a point of view of preventive medicine. We cannot, however, detect insecticide levels in the range of those which occur in the blood of the general population with the only exception of pp'DDE.

During 1963 52 blood samples of our insecticide workers were examined in Tunstall Laboratory by this method. From 1964 onwards we used this analytical method in the chemical laboratory of our companies at Pernis. It became a routine procedure in the periodic medical examination of insecticide workers and soon the results proved to be the single most important data to present from a preventive point of view.

At the beginning in 1964 our detection levels for HEOD (dieldrin), endrin and Telodrin respectively were 0.01, 0.01 and 0.004 μg/ml, since then they have been lowered to 0.005, 0.005 and 0.002 μg/ml respectively. From April 1969 on pp'DDE is also measured with a detection level of 0.005 μg/ml.

The technique used and the results obtained from the insecticide determination in the blood of our workers will be discussed in chapter 13.

2.3.3. *Industrial safety levels; dieldrin equivalent*

By comparing blood levels with clinical, biochemical and electroencephalographic data, it soon became apparent that the threshold level in the blood below which no signs or symptoms of intoxication occur, is at or above 0.20 μg/ml for dieldrin and 0.015 μg/ml for Telodrin. The threshold level concept is used here in the sense that Kehoe (1965) used it for lead.

As our experience with measurable endrin levels is only small — at the level of sensitivity of our analytical method — we are unable to establish a definite threshold level in the blood for endrin intoxication. Available data, however, together with extrapolation from animal data, as will be discussed in chapters 6 and 7, suggest that in practice the threshold level for endrin intoxication may be used as lying between 0.050 and 0.100 $\mu g/ml$ endrin in the blood.

Our figure for dieldrin is not in disagreement with the threshold level previously suggested by Brown *et al.* (1964) at 0.15 $\mu g/ml$, since we understand that their suggested threshold level is extrapolated from blood levels in samples derived 2-5 weeks after intoxication took place and insufficient allowance could be made for the movements of dieldrin in the different types of intoxication as to varying acuteness (see chapter 16 on "Patterns of intoxication").

We have used the above mentioned threshold levels for dieldrin and Telodrin as industrial safety levels since 1964, consequently every worker in which we find insecticide levels exceeding these thresholds is transferred to other work.

A practical problem in connection with the application of these safety levels in our insecticide plant was the fact that in 1964 and 1965 many of our workers were exposed to aldrin, dieldrin and endrin as well as to Telodrin, simultaneously in the formulation unit, or successively in the production units. To overcome this practical problem we introduced the concept "Dieldrin Equivalent" based on the estimation that the safety limit for the Telodrin level in the blood is at least 10 times lower than that for dieldrin. So in expressing the total insecticide level in the blood of our workers with mixed exposures we calculated:

$D + 10\ T = D.Eq.$

(dieldrin level in $\mu g/ml$ + 10x Telodrin level in $\mu g/ml$ = Dieldrin Equivalent in $\mu g/ml$), expressed as a dieldrin level.

Again endrin was not included in this conversion formula, as under normal working conditions endrin is never detected in the blood of our insecticide workers at our detection levels.

After having applied this concept of dieldrin equivalent for more than 6 years we still find it very satisfactory. We now believe that originally the effect of the Telodrin was underestimated, but as Telodrin is not handled any more this has no practical consequences. It only means now that in mixed exposures, levels expressed in D.Eq. probably are under-valued on account of an underestimation of the Telodrin part of the exposure.

Although Telodrin production was discontinued in September 1965, we continued using the term D.Eq. to compare blood levels of those workers in whom we still find Telodrin as well as dieldrin in the blood. The fact that the chemically active sites of the molecules of the insecticides under discussion are very similar – as described in chapter 3 (Soloway 1965) – justifies to a certain extent this simple addition. According to Varley (1968) another justification is that the therapeutic equivalence of drugs is a more important criterion than chemical or availability equivalence from a pharmacological point of view. This means, that the ideal criterion for the establishment of the therapeutic (or in the same way toxicological) equivalence of a drug (or a toxicant) is the comparison of the efficacy in clinical application. As aldrin, dieldrin, endrin and Telodrin have the same effector organ, the central nervous system, and intoxications with these insecticides clinically resemble each other so closely, we feel that the continued use of the D.Eq. concept is justified.

Apart from the preventive value of transferring over-exposed workers, the determination of insecticide levels in the blood allowed us to detect the danger spots in the plants, to advise technical improvements to be made there and/or better precautions to be taken. It also helped to identify careless workers, give them better instruction and supervision, or transfer them from the job. All this resulted in an improved hygienic condition in our insecticide plant, which finds its expression — as will be seen in chapter 13 — in a steady decrease of the average blood level of workers in this plant between 1964 and 1970. In fact our industrial safety levels will now only be reached by accidental gross over-exposure.

2.4   GENERAL HEALTH SURVEY

2.4.1 Insecticide intoxications

The group of workers who have had an insecticide intoxication in the past are of interest since they must represent an extreme exposure group even though insecticide levels in their blood at the time of intoxication are not known. These workers must have had the highest insecticide exposures of all workers described, therefore should this be the case they should have been the first to show signs of damage to tissues or organs or signs of disturbance of body functions.

As was discussed earlier, intoxications caused by aldrin, dieldrin, endrin or Telodrin cannot be distinguished from each other on clinical signs and symptoms. However, different types of intoxication may be distinguished, as was done by Hayes (1963) for dieldrin: "Three syndromes (determined largely by the size and number of doses) may be recognized:

1. A few large doses produce increasing stimulation of the central nervous system culminating, if the dosage is sufficiently high, in one or more convulsions. If death does not occur, there is relatively prompt recovery without significant weight loss or other permanent injury.

2. A larger number of moderate sized doses may produce without warning a condition marked by complete loss of appetite, weight loss, and convulsions. Without treatment death is apparently inevitable.

3. Many relatively small doses may produce one or a few convulsions with lesser accompanying symptoms that may recur even though exposure is discontinued".

As regards this latter suggestion of recurring convulsive symptoms after disconti-nuation of exposure, one should realize that in developing countries, with consequently developing health services, the causes for both convulsive disorders and conditioned convulsive responses to stresses are many and manifold. Objective medical recording of delayed convulsive responses to these insecticides has not been reported. The occurrence would be a contradiction with both our own not inextensive medical experience and also with laboratory experiments with various species of animals. Therefore we firmly believe that a cause other than delayed manifestations of intoxication should be sought.

On account of the description in the literature of cases of intoxication, our own experience and the data on insecticide levels in intoxication cases, we agree that 3 types of intoxication may be distinguished. These three types, however are not quite identical with those described by Hayes, as will be described in chapter 16.

2.4.2 Current general health

A review of the medical histories of our insecticide plant employees with longstanding high insecticide exposure as well as of those with manifest intoxication, is of importance. In this context special attention should be paid to the central nervous system as the effector organ and to other organs with a high fat content in which these lipophilic substances tend to accumulate: the liver, the kidneys and the bone-marrow.

2.4.3 Some health parameters

In addition to a review of the medical histories of the workers special consideration will be given to certain health parameters, such as:

a. *Body weight*: a non-intentional decrease in body weight is generally regarded as an expression of a negative influence on general health. An increase does not necessarily mean the opposite.

b. *Blood pressure*: a slight increase in blood pressure may be expected with ageing. Anything abnormal might point to cardiovascular or renal disease or to a disturbance in the endocrine system.

c. *Erythrocyte sedimentation rate (S.R.E.)* as another aspecific parameter for general health. According to Pincherle *et al.* (1967) values below 12 mm after 1 hour may be regarded within the normal range. According to these authors the S.R.E. increases with age, smoking habits, obesity and independently with hypercholesterolaemia. It also undergoes seasonal fluctuations.

d. In order to disclose possible reactions from *blood* and *blood-forming tissues* at early onset the following measurements were included: hemoglobin, RBC's, hematocrit, MCV and WBC's. Differential white cell counts were done during the earlier production years but were subsequently discontinued as they always proved to be within normal ranges.

e. *Rate of absenteeism due to disease and accidents* is another indicator of general health when compared with the same data from a comparable group of unexposed workers.

2.5 OTHER NON-SELECTIVE PARAMETERS

2.5.1 Introduction

During the pre-clinical or pharmacotoxical stage of intoxication an effect on certain organ functions takes place, a compensation of which might be measured in non-selective parameters before subjective or objective signs or symptoms other than those expressed in these specific parameters occur (De Bruin and Zielhuis 1967, Hatch 1968).

As breakdown of absorbed foreign substances takes place mainly in the liver, biochemical changes associated with liver function, and to a lesser extent changes in blood and renal parameters, may be expected first after absorption.

In screening for stress on, or disturbances of, the liver function a whole range of liver function tests is available, most of which, however, are far too insensitive for this sort of preventive screening. There is no need to discuss here all these liver function tests, only the parameters used in this study will be described and the reason for using them. Liver function tests suitable in screening for early responses to a toxicant or other stresses may be divided into:

a. *Structural injury tests*, which give an indication of the integrity of the liver cell membrane and for incipient or occurring injury to this structure. As such we used SGOT, SGPT and LDH.

b. *Functional response tests* to hyperphysiological and stress situations. The latter include biochemical, sub-cellular and electronmicroscopic reactions to hyperphysiological exposures to homologous or foreign substances. In this category changes in total serum protein and the serum protein electrophoresis pattern were used, as well as serum alkaline phosphatase levels and enzyme induction tests.

Collectively, these tests constitute a good early warning system of liver stress, disturbances in liver function and possible injury to liver tissue notably so if groups of exposed workers are considered rather than individuals.

2.5.2 Structural injury tests

The principle of determination of enzyme levels in blood serum is that body cells allow the passage of enzyme to the extracellular body fluids as soon as there is an increase of permeability of the cell membrane. This need not mean cell necrosis. Already submicroscopic changes in the cell membrane – without changes detectable by electronmicroscope – may cause this effect. For most enzymes there is a high concentration gradient between liver cells and blood plasma. The integrity of the cell membrane depends on these gradients at the cost of much energy. All processes which stress the energy production of the liver cell may in this way disturb the integrity of the cell membrane, resulting in increased enzyme levels in the blood serum. Particularly hepatotoxic chemicals tend to consume much energy in the detoxication process (Zondag 1963, De Bruin and Zielhuis 1967-II). A similar course of events may occur during hyperthermia, radiation and physical stress.

Although the occurrence of most enzymes is not restricted to one tissue or organ in the body, a combination of different enzyme tests or a sub-division into isoenzymes enables the delineation of the organ or tissue affected since different body tissues display different and typical "enzyme patterns". Thus in the case of liver stress or damage caused by hepatotoxic compounds or by intra- or extrahepatic obstruction of the bile flow a simultaneous rise of the SGOT, SGPT and LDH activities in the blood serum should be expected. The LDH levels may be normal again within a few days; however, it may take weeks before the SGOT and SGPT levels have returned to normal. Initially the de-Ritis quotient (= SGOT/SGPT) falls below 1, later returning to a normal slightly larger than 1 (about 1.1) ratio. The reason is that with membrane disturbance only more GPT than GOT will pass from the cell because GPT is a purely cytoplasmatic enzyme and GOT is 50% cytoplasmatic and 50% mitochondrial. So if SGOT rises significantly above SGPT this means necrosis of the liver cell, for only then can the mitochondrial GOT be released.

These considerations make it possible to differentiate between the degrees of the liver disturbance:

SGOT and SGPT increased; SGOT/SGPT>1: a sign of liver cell injury or necrosis with morphological change of liver cells.

SGOT and SGPT increased; SGOT/SGPT<1 (about 0.8): a sign of liver cell stress with increased membrane permeability but without liver cell necrosis and without morphological changes of liver cells.

SGOT and SGPT normal; SGOT/SGPT>1 (about 1.1): normal state of affairs (Zondag 1963, De Bruin and Zielhuis 1967-II).

As long as membrane function is intact, however, a normal enzyme level in serum is no proof that there is no change in intracellular enzyme activity.

In some perfectly healthy people SGPT may be higher than SGOT. In cirrhosis of the liver both tend to be increased to a certain extent with SGOT/SGPT>1 (Schalm 1969).

A serum LDH activity up to 300 mU/ml is considered to be normal. LDH is a normal constituent of many body tissues, therefore not being a very selective indicator for the organ involved. Increases of the LDH activity return quickly to the normal, sometimes within one or two days (Zondag 1963).

2.5.3 Total serum protein and serum protein spectrum

The technique of serum protein electrophoresis has shown that the serum protein pattern is a sensitive indicator particularly for disturbances in the function of the liver. Many toxic compounds may cause significant changes in the albumine-globuline ratio at an early stage of intoxication (De Bruin and Zielhuis 1967-III). The primary function of albumin is to maintain the colloid osmotic pressure of the plasma, which among other things means the maintenance of a constant blood volume. Albumin synthesis takes place in the liver: exposure to hepatotoxic compounds may decrease the albumin fraction in the blood plasma.

α_1- and α_2-globulins contain sugar- and fat-bound proteins, amongst others, compounds which play a role in the aspecific defence against infections and other harmful influences. For their synthesis they too depend to a large extent on the liver.

β-globulins consist of sugar- but foremost of fat-bound proteins which have a function as transport proteins for hormones, phospholipids, vitamins, iron (PBI) and many drugs and chemicals. The main site of synthesis is again the liver.

α_1-, α_2- and β-globulins may all be increased, or only one of them, as a sign of adaptation to the presence of invading micro-organisms or absorption of toxicants. In the case of an increase in β-globulin this may be a sign of increased transport facilities. Depression of α- and β-globulins is a sign of insufficient synthesis of these fractions caused by disturbed liver function.

The γ-globulin fraction contains all kinds of antibodies against antigenes of bacteria, viruses, toxines, foreign proteins or sometimes auto-proteins. Synthesis of this fraction takes place in the reticulo-endothelial system, especially in plasma cells derived from small lymphocytes. γ-globulins may be increased relatively and/or absolutely in conditions of disturbed liver function.

In liver cirrhosis albumin depression is a constant sign. γ-globulin may be more or less increased, α-globulin normal or subnormal, β-globulins are sometimes increased. (De Bruin and Zielhuis 1967-III, Van Dommelen 1961).

The thymolturbidity test is one of a series of turbidity tests used as a crude indicator for the albumin-globulin ratio. As such, an abnormal TTT is a sign of disturbed liver function.

2.5.4 Serum Alkaline Phosphatase

Alkaline phosphatase is an enzyme formed in the osteoblasts and has a function in the building and repair of bone. Serum alkaline phosphatase is increased in children up to puberty and in many bone diseases. Apart from this, alkaline phosphatase may be increased in the blood serum in certain hepatic disturbances. Alkaline phosphatase is excreted by the liver, and some alkaline phosphatase may be synthesized in the liver as well. The alkaline phosphatase level in blood serum will increase with the slightest increase in pressure, for whichever reason, in the bile excreting system, be this an intra- or extrahepatic obstruction, such as a calculus, a tumour or metastasis,   or fibrosis, as well as intrahepatic cholestasis due to drug action. The first slight increases occur during the adaptive phase, accompanied by a slight increase of the liver volume, and probably may be regarded as an indirect indication of enzyme induction. (Fujimoto *et al.* 1965, Walker 1966, 1967, Wright *et al.* 1967, 1968).

Thus the alkaline phosphatase in blood serum may be raised before SGOT, SGPT and LDH are increased, and again serum alkaline phosphatase may still be increased when SGOT, SGPT and LDH have returned to normal. The serum alkaline phosphatase level may be regarded as a sensitive indicator of response in a phase before structural injury occurs.

2.5.5 Enzyme induction tests

The rapidly increasing volume of literature on enzyme induction and interference with metabolism of drugs and/or chemicals in animal experiments is in marked contrast with the scarcity of data available on practical studies in humans on the same subject. In principle all these tests involve the measurement of the duration of the action of standard drugs or the amount of these drugs or metabolites excreted over a certain time, as compared with known data or with a control group. Hexobarbital and zoxazolamine are useful drugs for this purpose. Phenylbutazone may also be used.

Some of the more easily measurable effects of the stimulation of the drug metabolizing enzyme system is a shortening of the hexobarbital sleeping time and an increase in the bromosulphthalein excretion (Fujimoto *et al.* 1965).

Drugs and chemicals may stimulate the hydroxylation of steroids in the body. The measurement of the urinary excretion of the metabolite of cortisol: 6-β-hydroxycortisol as compared with total 17-hydroxycorticosteroids (which are not changed by the inducers) may be a useful index for the induction of hydroxylases in liver microsomes in man (Conney 1967, Kuntzman *et al.* 1968).

In animal experiments the BSP test is used in what we might call a reversed way, *i.e.* to screen for an accelerated BSP clearance as a result of BSP metabolism enhanced by enzyme inducers. In normal clinical use this bromosulphthalein test is used to screen for decreased clearance as a measure for impaired liver function. We do not regard this a suitable test for routine use in preventive medicine as it involves, apart from taking at least two separate blood samples, the injection of a foreign substance to which adverse reactions may occur, and have occurred in actual clinical practice.

Most of these tests, even the steroid measurements in urine are not suitable for routine preventive practice.

Serum alkaline phosphatase, as was discussed in 2.5.4, may be an early indicator for enzyme induction.

Another proposition seems to be the determination of d-glucaric acid in urine, which, according to Aarts (1965) may prove to be a useful test for enzyme induction in man, as typical inducers cause an increased excretion of d-glucaric acid in the urine.

But of course everything depends on what microsomal enzyme systems are stimulated by a certain inducer.

For the cyclodiene insecticides, perhaps the most reliable test in man will be a comparison of pp'DDE levels and cholinesterase activity in the blood of workers exposed to cyclodiene insecticides (but not to DDT), with a non-exposed control group. In experimental animals comparatively high dosages of the cyclodiene insecticides are known to influence these parameters. To cite a study of Davies *et al.* (1969-a): "A study of blood levels of pp'DDE (the principal metabolite of DDT) in a group of patients taking anti-convulsant drugs for more than three weeks, has revealed strikingly lower DDE levels than have been found in the general population. Diphenylhydantoin appeared to have more potent action in reducing DDE blood levels than phenobarbitone". The same finding was made on samples of adipose tissue.

2.6 SUMMARY AND CONCLUSION

In this chapter those toxicological concepts which are used in this study were briefly discussed. In doing so the controversial choice between "no-effect level" and "no-adverse-effect level" was mentioned, and our reason for preferring the term "no-adverse-effect level" was given.

Enzyme induction was discussed to some length and we may conclude, that at this time there is no proof that enzyme induction as such is necessarily an adverse effect, although it may be a measurable effect.

As the method of choice for determining total exposure to organochlorine insecticides, the determination of the blood levels of these compounds by GLC with electron capture detection was established.

Parameters used in this study with the purpose of demonstrating the existence or non-existence of responses to the exposures measured were discussed.

As was posed in the first part of this chapter it will then be possible to compare exposure as determined by biological monitoring against the responses found in selective and non-selective parameters and the general health survey. The result shows the interaction of total exposure and individual disposition (De Bruin and Zielhuis 1967-I, II, III, IV).

And, to conclude the general introduction to this study with the words of Hayes (1967): "Toxicologists have just as much responsibility to advise of safety as they have to warn of danger".

Chapter 3

The insecticides aldrin, dieldrin, endrin and Telodrin – general information

3.1 Introduction
3.2 Common names and chemical identities
3.3 Comparison of chemical structure
3.4 Physico-chemical characteristics of technical products
3.5 Commercial compounds
3.6 Summary

3.1   INTRODUCTION

Aldrin, dieldrin, endrin and Telodrin (isobenzan) with heptachlor, chlordane, etc. form the group of cyclodiene insecticides. They are a subgroup of the chlorinated cyclic hydrocarbon insecticides to which DDT, BHC, toxaphene, etc. also belong, nowadays generally called organochlorine insecticides.

Cyclodiene insecticides are cyclic hydrocarbons having a chlorine substituted methanobridged structure. The four insecticides under discussion are manufactured from hexachlorocyclopentadiene by way of the Diels-Alder-diene reaction.

Aldrin, dieldrin and endrin in various commercial formulations have many end-uses varying from public health schemes to crop protection and industry. Major uses are:

Aldrin: a broadspectrum insecticide primarily used for the control of a wide range of soil pests, grasshoppers and certain cotton insects.

Dieldrin: a broadspectrum insecticide to control certain insects attacking principal field-, vegetable- and fruitcrops. It is also widely used against public health pests including disease vectors and in locust and termite control.

Endrin: the control of a wide range of foliage pests of cotton, rice, tobacco, maize, sugar cane and fruit trees, such as cutworms, armyworms, aphids, corn borers, cabbage looper, grasshoppers, plant bugs, webworms and many other pests. Endrin is particularly effective against caterpillars.

Aldrin, dieldrin and endrin are Shell-products, manufactured in Holland and in the U.S.A. Endrin is also manufactured by Velsicol in the U.S.A.

Telodrin was manufactured by Shell in Holland. Production was discontinued in September 1965.

3.2   COMMON NAMES AND CHEMICAL IDENTITIES

The chemical identities and the various names used for these insecticides are given below and other relevant data are given in Table 2. Isodrin belongs to this group but has never been marketed.

Aldrin is the common name for a technical insecticide which contains at least 95% HHDN, the abbreviation for 1, 2, 3, 4, 10, 10-hexachloro- 1, 4, 4a, 5, 8, 8a-hexahydro- 1, 4, 5, 8-endo-exo-dimethanonaphthalene. Aldrin is the endo-exo stereoisomer of isodrin and is an intermediate in the production of dieldrin.

44

Isodrin is the common name for an intermediate in the production of endrin. It is the endo-endo stereoisomer of aldrin.

Dieldrin is the common name for a technical insecticide which contains at least 85% HEOD, the abbreviation for 1,2,3,4,10,10-hexachloro-6,7-epoxy-1,4,4a,5,6,7,8,8a-octa-hydro-1,4,5,8-endo-exo-dimethanonaphthalene. Dieldrin is the epoxide of aldrin and is the endo-exo stereoisomer of endrin.

Endrin is the common name for a technical insecticide which contains at least 92% of the endo-endo stereoisomer of dieldrin. Endrin is the epoxide of isodrin.

Telodrin is the proprietary name for the technical insecticide isobenzan which contains at least 95% 1,3,4,5,6,7,8,8-octachloro-1,3,3a,4,7,7a-hexahydro-4,7-methanoiso-benzofurane.

3.3   COMPARISON OF CHEMICAL STRUCTURE

In Fig. 5 the structural formulas and the spatial configurations of these insecticides are shown diagrammatically, although actually the spatial configuration is nearly spherical.

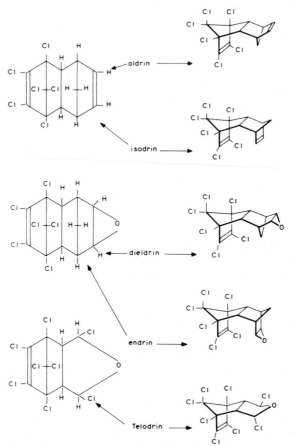

Fig. 5. Chemical structure and spatial configuration.

The chlorinated "left" side, which represents almost 75% of the mass of the molecule is identical in all five chemicals (Fig. 6).

Fig. 6. Common structure of discussed insecticides.

Soloway (1965) examined the insecticidal action of 106 cyclodienes and found a high insecticidal activity only in those compounds in which there are two electronegative centres, placed close to each other and situated in the plane of symmetry as defined by the dimethanobridge. This means, that a projection on this plane of symmetry of the molecular model must fall within a certain critical outline (Soloway 1965).

The molecular outlines of the four insecticides under discussion are given in Fig. 7. The first electronegative centre – the hexachlorinated part of the molecule – predominates on the left. On the lower right is the second electronegative centre consisting of oxygen and/or chlorine atoms. In fact, there is remarkably little difference in the outlines of these four insecticides.

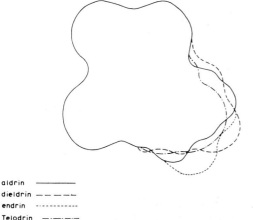

aldrin ——————
dieldrin — — — — —
endrin ···············
Telodrin —·—·—·—

Fig. 7. Outlines of the projection on the plane of symmetry of compact molecular models of aldrin, dieldrin, endrin and Telodrin (modified after Soloway 1965).

These insecticides have similar effects on the central nervous system and it is likely that they act on similar sites within that system. However, as the electron-rich sites are not considered to be chemically active (Soloway 1965, Benson 1969), Soloway presumed that they are likely locations for strong electrostatic interactions. It therefore appears, that physical interactions rather than chemical reactions occur between the electronegative centres of the insecticides and an active biological site – possibly on the nerve cell membrane – but only when a ready fit of the two occurs as determined by the critical outline of the molecule.

Further results of this study by Soloway suggest, that the epoxides are the active compounds, while only those other compounds in which the double bond on the non-chlorinated side of the molecule is readily epoxidized, display insecticidal activity.

Hathway (1965) considered that the rigid-case-configuration with the electro-negative centres of the chlorine atoms of these lipid-soluble insecticides was important to both the passage of these substances across membranes and to their molecular action.

It is in line with these similarities in chemical configuration that, despite some differences in their physiological effectiveness on a unit weight basis, their toxicological and clinical effects are very similar in most mammals and man. Many authors agree on this point.

3.4 PHYSICO-CHEMICAL CHARACTERISTICS OF TECHNICAL PRODUCTS

Some physico-chemical characteristics of the technical insecticides are given in Table 2 and may be of assistance in estimating the relative hazard of each compound.

Table 2
PHYSICO-CHEMICAL CHARACTERISTICS OF TECHNICAL INSECTICIDES

	Aldrin	Dieldrin	Endrin	Telodrin
General aspect	whitish to light-brown crystalline			
Empirical formula (pure compound)	$C_{12}H_8Cl_6$	$C_{12}H_8Cl_6O$	$C_{12}H_8Cl_6O$	$C_9H_4Cl_8O$
Molecular weight (pure compound)	365	381	381	412
Spec. gravity at $20^\circ C$	1.70	1.70	1.70	1.87
Melting point $^\circ C$	102	150	235	120
Flashpoint	non-flammable and non-explosive			
Solubility	practically insoluble in water; moderately soluble in most organic solvents; highly soluble in lipid material			
Vapour pressure at $25^\circ C$ in mm Hg.	6×10^{-6}	1.8×10^{-7}	2×10^{-7}	1×10^{-5}
TLV in air mg/m^3	0.25	0.25	0.10	not considered

The threshold limit values (TLV's) as recommended by the American Conference of Governmental and Industrial Hygienists (ACGIH) (Threshold Limit Values, 1969) have also been included. The ACGIH defines the TLV as follows: "The Threshold Limit Values refer to air-borne concentrations of substances and represent conditions under which it is believed that nearly all workers may be repeatedly exposed, day after day, without adverse effect. Because of wide variation in individual susceptibility, exposure of an occasional individual at or even below the threshold limit may not prevent discomfort, aggravation of a pre-existing condition, or occupational illness".

Vapour pressures of aldrin, dieldrin and endrin are so low, that at normal temperatures air saturated with the vapour of these compounds contains concentrations lower than the recommended TLV's. The crystalline state results in a lower inhalation hazard as compared with powder or dust under the same conditions of work.

3.5   COMMERCIAL COMPOUNDS

Technical insecticides as such are handled only during manufacture and in formulation plants. The bulk of the production is transported as such to formulation plants overseas or in Europe, while the remainder is formulated in our own formulation unit.

Technical insecticides cannot be used as such in agriculture since it is impossible to distribute them on soil, crops, etc. uniformly and efficiently in the low dosages which are required to meet the biological, toxicological and economic requirements.

The production of commercial compounds ready to use in the field is a process called formulation in the chemical industry. In this formulation process the technical product is mixed, and thus diluted, with fillers or solvents. In some instances, *e.g.* dusts, the resultant compound is milled, and in others, *e.g.* wettable powders and emulsifiable concentrates, emulsifiers, stabilizers, etc. are added in small quantities to facilitate further dilution with water.

The most important groups of commercial products are:

Wettable powders (w.p.'s) containing from 25-50% of the pure insecticide. They have to be diluted with water before use in the field.

Dust concentrates (d.c.'s) containing from 20-75% of the pure insecticide in the case of aldrin and dieldrin or 10% for endrin. They have to be diluted with inert filler or sand before use in the field, or sometimes they can be used as such.

Field strength dusts (f.s.d.'s) containing 1¼-5% of the pure insecticide with aldrin and dieldrin, ½-1% with endrin. Ready for use in the field.

Emulsifiable concentrates (e.c.'s) containing 15-48% of the pure insecticide. They have to be diluted with water before use in the field.

Granules contain 10-20% aldrin or dieldrin or 2-5% endrin. Granules are ready for use in the field.

Apart from these more common groups there are other products such as special formulations for locust-control, combinations with organophosphates and so on. In all cases the concentration percentages and the names of the constituent insecticide(s) are mentioned on the label together with the correct method for further dilution and handling of the product, the safety precautions to be taken and first aid instructions.

In powders, dusts and granules inert mineral fillers are used such as Fuller's earth, limestone, china clay, etc. In the field use of dust formulations the inhalation hazard is relatively higher than with products such as granules.

In emulsifiable concentrates high-boiling high-aromatic solvents are used, such as:
Shellsol-A: boiling range 161-181°C; flashpoint 45°C; spec. gravity 0.875. and
Xylene: boiling range 137-143°C; flashpoint 24°C; spec. gravity 0.860.
In handling e.c.'s the flammability must be considered, as well as the fact that the solvent used may facilitate absorption of the toxicant through the skin.

The spray-mist created when applying diluted w.p.'s or e.c.'s form an inhalation hazard comparable with that from dust application.

It is outside the scope of this study to discuss toxicological data of all components of the compound products; only those facts relevant for the estimation of the hazard of the compound have been mentioned.

48

3.6 SUMMARY

Common and chemical names and the relevant physico-chemical properties were given for the four insecticides under discussion. The principal differences in the compounds themselves, their main uses and hazards in use, were described.

The chemical structures of these cyclodienes were compared and from this it may be concluded that all four have essentially a very similar molecular profile and electrostatic configuration. From this — and from the similarity of clinical effects — it is likely, that all four exercise their toxicological action on a similar, and probably the same, site in the central nervous system.

B. SURVEY OF LITERATURE

Chapter 4

General toxicology of aldrin and dieldrin — animal studies

4.1 INTRODUCTION

In this review of the literature on aldrin and dieldrin only recent information will be discussed. Reference will be made to earlier studies only when necessary to present the newer data in perspective. These earlier studies have been well summarized in various reports, such as those of the FAO/WHO Joint Meeting (1965, 1967), Van Genderen (1965) and Hodge *et al.* (1967).

Since reproduction and teratogenesis are very probably irrelevant as regards the long-term industrial exposure of male workers, reference to the extensive literature on these subjects has been deliberately omitted: not only in the case of aldrin and dieldrin, but also for endrin and Telodrin.

In this study aldrin and dieldrin are considered together, because aldrin is rapidly epoxidized to dieldrin in the animal and human body (Treon *et al.* 1955, Bann *et al.* 1956, Wong *et al.* 1965).

4.2 ACUTE AND SUBACUTE TOXICITY

A summary of the more important acute toxicity data from animal tests is given in Table 3. Many more data on the acute toxicity of aldrin and dieldrin are tabulated by Hodge *et al.* (1967).

From these data, as already mentioned by Treon *et al.* (1955) and Heath *et al.* (1964), it appears that aldrin and dieldrin are slightly more toxic to female than to male rats.

The LD_{50} varies with the concentration (Barnes *et al.* 1964) and the vehicle employed (Heath *et al.* 1964). Organic solvents and vegetable oils increase the toxicity due to the enhanced rate of absorption of the toxicant into the body.

The toxicity also varies between species, although in the species mentioned in Table 3 the variations are not very marked.

52

Table 3

ACUTE TOXICITY OF ALDRIN AND DIELDRIN IN EXPERIMENTAL ANIMALS

Animal species			Route of administration	Formulation	LD_{50} in mg/kg body weight		References
					aldrin	dieldrin	
Rat -	WS	♀	Oral	Arachis oil	–	50.8	Barnes et al. 1964
"	"	♂	"	"	–	63.5	Heath et al. 1964
"	CFE		"	Peanut oil	45.9	38.3	Treon et al. 1955
"	"		"	Arom. solv.	18.8	–	"
"	"		"	40% e.c.	56.4	–	Muir 1968
"	"		"	40% w.p.	62.5	–	"
"	"		"	75% d.c.	72.2	–	"
"	"		"	2½% f.s.d.	109.0	–	"
"	"		"	20% e.c.	–	55.9	"
"	"		"	50% w.p.	–	52.1	"
Mouse			"	–	95	75-100	Korte et al. 1963
"			"	–	44	38	Borgmann et al, 1952
Guinea pig			"	–	33	49-59	Borgmann et al. 1952
Rabbit			"	–	50-80	45-50	"
Dog			"	–	65-95	56-80	"
Sheep			"	–	–	50-75	"
Rat -	CFE		Dermal	40% e.c.	194.0	–	Muir 1968
"	"		"	40% w.p.	274.0	–	"
"	"		"	75% d.c.	269.0	–	"
"	"		"	2½% f.s.d.	>100.0	–	"
"	"	♂	"	20% e.c.	–	213.8	"
"	"	♀	"	20% e.c.	–	119.9	"
"	"		"	50% w.p.	–	213.4	"
Rat -	WS	♀	Intraperit.	Glycerol-formal.	–	55.9	Heath et al. 1964
"	"	♀	Intravenous	"	–	8.9	"
Mouse			Intravenous	–	21.5	15.2	Korte et al. 1963

Borgmann et al. (1952) cited after FAO/WHO 1965 and 1967

WS = Wistar strain   CFE = Carworth Farm E strain
e.c. = emulsifiable concentrate   w.p. = wettable powder
d.c. = dust concentrate   f.s.d.= field strength dust

The largest variations in toxicity occur between the different routes of entry: highest toxicity via the intravenous route, lower via the oral route and lowest from dermal application. The explanation for this is, that intravenous and intraperitoneal injections rapidly produce high concentrations in the blood and central nervous system, whereas by the other routes resorption is slower and blood concentrations remain lower for the same given dose because during absorption some of the toxicant is stored in adipose tissue and some metabolized and eliminated. Thus higher oral and dermal doses are required to achieve those concentrations in the central nervous system necessary for the production of signs of intoxication.

The symptoms of aldrin and dieldrin intoxication are similar to those of intoxication caused by endrin, Telodrin and other cyclodiene insecticides. Although there are minor variations between the effects of these insecticides in different species, it is generally impossible to distinguish between the toxicants on the grounds of symptomatology alone (Brown *et al.* 1962).

The main symptoms of acute intoxication are increased irritability, tremor, followed later by tonic-clonic convulsions, which indicate that the principal site of action is the central nervous system. The onset of symptoms depends mainly on the dose employed and the route of administration. In the rat after intravenous injection of a toxic dose of aldrin or dieldrin epileptiform convulsions occur within 2-5 minutes; while those following oral dosing appear after about 40 minutes. Doses in the lethal range usually cause death after the 4th-6th convulsion, but those rats, which survived non-lethal doses recovered fully and showed no delayed effects (Heath *et al.* 1964).

The same authors found, that rats underfed for a prolonged period before dosing, were considerably more susceptible to toxic doses of aldrin and dieldrin than those fed normally, probably due to lower storage capacity in adipose tissue and consequently higher insecticide levels in the blood and in the central nervous system. In their conclusion they state: "The evidence, therefore, is that in the rat and in man dieldrin does not produce a long-lasting lesion in the central nervous system. The toxic effects can be accounted for by the dieldrin metabolized in the body at the time they take place, and they disappear with the dieldrin".

Brown *et al.* (1964) fed 4 beagles dieldrin at a daily oral dose of 0.4-0.8 mg/kg body weight. After 8 episodes of convulsions the HEOD (dieldrin) levels in the blood of the dogs ranged from 0.27-1.27 μg/ml.

Keane *et al.* (1969) fed 18 mature mongrel dogs at various daily doses of dieldrin in corn oil until intoxication occurred. A direct relationship was established between the dieldrin concentration in the blood and the severity of clinical signs of intoxication: on the day of the first muscular spasms the average concentration was 0.38-0.50 μg/ml and at the time of the first fully recognisable convulsion 0.74-0.84 μg/ml.

4.3 SUBCHRONIC AND CHRONIC TOXICITY

4.3.1 Studies in rats

Until recently the evaluation of the chronic toxicity of aldrin and dieldrin was essentially based on two studies.

Treon *et al.* (1955) fed male and female rats diets containing 1.5, 12.5 and 25 ppm * of either aldrin, dieldrin or DDT for two years. A possible slight increase in the liver-weight/body-weight ratio was observed in both males and females fed either of the three organochlorines at all levels. Non-specific changes in hepatic cells were recorded in

---

*Most reports only express dietary concentrations in ppm, and do not mention the conversion factor into mg/kg body weight. This conversion depends on, amongst other things, the type of diet, particularly the moisture content. Concentrations in ppm have not been converted by us to mg/kg/day, unless given in the original report.

all experimental groups, but the changes occurred with greater frequency in the animals fed DDT.

The second feeding study was conducted by Fitzhugh et al. (1964). Male and female rats received diets containing 0.5, 2, 10, 50, 100 and 150 ppm aldrin or dieldrin for two years. No gross differences were observed between the control animals and those fed 0.5 and 2 ppm aldrin or dieldrin. At dietary levels of 10 ppm and less liver cell changes were rated as slight, while the difference between the control animals and the animals at 0.5 ppm may have been due to chance (a "slight" degree of histological liver cell changes was observed in only one rat fed dieldrin at this level). Nevertheless, the 0.5 ppm dietary level was assumed to be the lowest showing an effect.

Because an indisputable no-toxic-effect level had not been established, Walker et al. (1968) initiated a study at Tunstall Laboratory in 1966. Groups of C.F.E.-rats (40 males and 40 females per group) were fed dieldrin at dietary levels of 0.1, 1.0 and 10 ppm for two years. General health, body weight, food intake and mortality were unaffected at all dose levels. The only clinical effect observed at the highest dose level (10 ppm) was increased irritability and the occasional occurrence of convulsions on handling. These effects were observed after three months of feeding. Hematological data showed no difference between control and experimental rats. The levels of plasma alkaline phosphatase, plasma glutamic-pyruvate-transaminase, total serum proteins and serum urea were similar in the control and all treated groups. At autopsy an increase of liver weight and liver/body weight ratio was observed in female rats only at the 1.0 and 10.0 ppm feeding levels. Histological examination failed to show the presence of any changes. Only at 10 ppm was evidence found of the so-called "chlorinated hydrocarbon liver" changes, characteristically observed in rodents. All organs were normal at gross and microscopical examination. The tumour incidence was similar in both treated and control groups. (The investigators were careful to exclude extraneous materials which might affect the liver in a manner histologically similar to that caused by chlorinated hydrocarbons. The possibility of such effects has been mentioned by, amongst others, Kimbrough et al. (1968) for pyrethrum).

This study showed that the only effect observed at the 1.0 ppm level was an increase in liver weight in female rats and it was not associated with any histological change.

4.3.2 Studies in dogs

Two studies on the chronic toxicity of aldrin and dieldrin to dogs formed the basis of previous reviews. The first was conducted by Treon et al. (1955) and in their study of slightly more than fifteen month duration, groups of dogs fed either 1 or 3 ppm aldrin and 1 or 3 ppm dieldrin survived the entire test period. Increased liver weights were observed at the levels of both 1 and 3 ppm dieldrin and at the level of 3 ppm aldrin. Minor liver cell changes were seen in males and females fed aldrin at 3 ppm but none in dogs fed dieldrin.

In the two-year study conducted by Fitzhugh et al. (1964) dogs were fed aldrin or dieldrin at dosages of 0.2, 0.5, 1.0, 2.0, 5.0 and 10.0 mg/kg. A no-effect level of 0.2 mg/kg (equivalent to 8 ppm in the diet) was established for both compounds. Neither clinical nor histopathological abnormalities existed at this dose level.

Because Treon et al. had observed some increase in liver weight at the 1 and 3 ppm dieldrin level, further work was initiated both al Cornell University and at the Tunstall Laboratory.

At Tunstall Laboratory two dogs, one male and one female, have been on a continuing dietary regimen of 0.2 mg/kg of dieldrin per day since 1961 (Walker 1966, 1967). The only effects observed were increased serum alkaline phosphatase activity (in both animals) and increased bromosulphthalein clearance (in the male). The latter change is considered to be a sign of stimulation of microsomal enzyme activity (Fujimoto et al. 1965, Wright et al. 1968). Liver function tests have given normal or super-normal values.

A two-year oral exposure study of dogs to dieldrin at dosages of 0.005 and 0.05 mg/kg/day (equivalent to 0.1 and 1 ppm in the diet) was initiated in the same Laboratory in 1965. The results of this study, which have also been published in the above mentioned reports, can be summarized as follows:

Groups of 5 male and 5 female litter mate beagles received daily capsules containing doses of 0.005 and 0.05 mg/kg dieldrin in olive oil. Control dogs received capsules containing olive oil only.

General health, behaviour and body weight remained unaffected at all levels. Electroencephalographic recordings showed no difference between the control and the top dose groups. The results of all hematological studies and urinalysis were similar in all groups. In the highest dose group the physiological decrease with age in the serumalkaline phosphatase activity which occurred in the lower dose group and controls, did not appear. Consequently the alkaline phosphatase level in the high dosage animals, was higher, after the 18th week, than in the controls. No changes in the electrophoretic pattern were observed, though the total serum protein showed a slight reduction in the male top dose group. Therefore the latter observation was not considered to be toxicologically significant.

At autopsy the only significant finding was an increase in the liver/body weight ratio in the top dose groups. This increase was not associated with any histological anomaly. Histochemically, the fat content of the liver cell appeared normal. The relative increase in the serum alkaline phosphatase activity, found to be due to an increase in the normal adult dog isoenzyme, was considered to be related to the stimulation of the liver-processing-enzyme activity, caused by dieldrin administration (Wright et al. 1968), and will be discussed later on.

Neither by clinical, biochemical nor histological tests could any liver injury be demonstrated. All organs were normal on both gross and microscopic examination.

4.3.3 Studies in monkeys

A chronic toxicity study on rhesus monkeys has been in progress since August 1963 at the Kettering Laboratory of the University of Cincinnati. Zavon et al. (1967) reported on the first 36 months of this study and Wright et al. (1969) after 5.5-6 years exposure. Thirty-one monkeys divided into 6 groups were exposed to 0, 0.01, 0.10, 0.50, 1.0 and 5.0 ppm dieldrin in their food. The 5.0 ppm exposure was later reduced to 2.5 and 1.75 ppm in all but one animal, after 2 monkeys in this group had died. All animals were examined for changes in weight, blood count, urinalysis, liver function (total serum-protein, serum isocitric dehydrogenase and prothrombin time), dieldrin content of fat and blood and histology of the liver. At the end of the 36 months period no adverse effects had been observed in those monkeys fed dieldrin at levels in the diet below 2.5 ppm, being equivalent to 60 - 100 $\mu g/kg/day$ in this experiment. The livers of male monkeys fed dieldrin at 0.01, 0.1 and 0.5 ppm for 5.5-6 years were comparable to controls in all parameters measured in the study (liver DNA and RNA, liver protein, microsomal protein, liver succinic dehydrogenase, alkaline phosphatase, glucose-6-phosphatase and microsomal content of cytochrome P-450). The livers from animals fed dieldrin at 1.0, 1.75 and 5.0 ppm differed from controls in a slightly enhanced capacity of isolated washed liver microsomes to hydroxylate certain substrates (the 1.75 and 5.0 ppm groups only) and in an elevation of liver microsomal cytochrome P-450. Otherwise there was no microscopic, electronmicroscopic or chemical evidence indicative of proliferation of the smooth endoplasmatic reticulum or other changes in the liver which could be attributed to dieldrin.

4.4 CARCINOGENESIS

In a previous two-year feeding study (Treon *et al.* 1955) in rats fed aldrin, dieldrin or DDT at levels of 0, 2.5, 12.5 and 25 ppm the incidence of tumours in the experimental animals was similar to that in the controls. In a later review (Cleveland 1966) one of the authors concluded that "neither aldrin nor dieldrin induced any increase in the number of tumours" and..... "that aldrin and dieldrin are neither carcinogenic nor tumorigenic".

An experiment, conducted with C_3HeB/Fe mice fed aldrin or dieldrin at 10 ppm for two years, showed a statistically significant increase in hepatic tumours. The tumours produced were morphologically benign (Davis *et al.* 1962).

Another study conducted in rats fed dieldrin or aldrin at 0, 0.5, 2.0, 10, 50, 100 and 150 ppm in their diets for two years was reported as showing..." an overall increase in the number of tumours, particularly among rats on the lower dosage levels of aldrin and dieldrin" (Fitzhugh *et al.* 1964). The absence of a dose response relationship (18 tumours in 41 aldrin or dieldrin fed rats at 0.5 ppm, 15 tumours in 41 aldrin or dieldrin fed rats at 2 ppm and 16 tumours in 40 aldrin or dieldrin fed rats on 10 ppm) does not agree with the suggestion that the "apparent tumorigenic properties of dieldrin can be related to a general type of effect".

Barnes (1966) in discussing these rat and mouse studies stated that "it is reasonable to conclude that neither DDT, aldrin or dieldrin behaved like a typical liver carcinogen in the experiments just outlined, so that another cause of the difference in the incidence of liver tumours must be sought." Van Genderen (1965) arrived at a similar conclusion and referred in this context to an unpublished two-year feeding study on rats performed by Van Esch using a dose level of 75 ppm in which no generalized tumorigenic action could be demonstrated.

Aldrin was included in a study to determine the effects of the simultaneous feeding of four pesticides to rats (Deichmann *et al.* 1967). When fed at a level of 5 ppm for two years, the survival rate was comparable to the controls. The number of tumours found in the aldrin fed rats was equal to that in the controls, *i.e.* a total of 15 tumours in the controls and a total of 15 tumours in the aldrin fed rats. One tumour in the control group was considered to be malignant, while two in the aldrin group were also described as being malignant. Virtually the same incidence was observed throughout all test groups in this respect.

Directed primarily towards the incidence of liver tumours, the study disclosed no increase associated with aldrin, either individually, or in various combinations.

The absence of a carcinogenic effect of dieldrin upon rats was again confirmed in a recently concluded feeding study by Walker *et al.* (1968), in which dieldrin was fed to groups of CFE-rats of both sexes at dietary levels of 0, 0.1, 1.0 and 10.0 ppm for two years. The tumours appearing in the experimental group were no more numerous than those in the control group, nor were they of a different type or distribution.

Beagles, 5 males and 5 females in each group, were fed at oral dieldrin doses of 0, 0.005 and 0.05 mg/kg/day (equivalent to 0, 0.1 and 1.0 ppm dietary level) for two years and showed no tumours at autopsy (Walker *et al.* 1968).

A study on the life-time exposure of mice to dieldrin at dietary levels of 0.1, 1.0 and 10 ppm is in progress in Tunstall Laboratory. The preliminary results of this continuing study (Stevenson *et al.* 1967, Weinbren *et al.* 1968, Walker *et al.* 1968, Wright et al. 1969) are as follows:

The longevity of the 0.1 and 1.0 groups is similar to that of the controls. The life-span of the 10 ppm group is shortened. Liver changes were present in all treated groups. These changes, mainly discrete multiple nodules appeared in the 10 ppm group after 9 months, occasionally in the 1.0 ppm group after 18 months, and rarely in the 0.1 ppm animals after 21 months feeding. In appearance these lesions were different in histopathological aspects and in behaviour from the spontaneous hepatoma's which occurred at a low incidence in the treated mice and in the control group, and also different from the liver carcinoma's that could be induced by feeding butter yellow (dimethyl aminoazobenzene). The dieldrin induced lesions in these mice, which are similar to those induced by phenobarbitone, have no invasive growth, do not metastasise, could not be transplanted to other mice and did not have autonomous growth, moreover they tended to decrease in size when dieldrin administration was discontinued. Therefore these liver nodules cannot be considered malignant tumours.*

4.5 METABOLISM AND BIOCHEMISTRY

Experiments on rats with Cl^{36}-labelled dieldrin show that intestinal absorption starts almost immediately following oral administration but the rapidity and extent of the absorption may vary with the vehicle used. The absorbed dieldrin is mainly transported by the portal vein blood, and only a small proportion via the lymph. Initially dieldrin is distributed most widely in the body, but redistribution takes place rapidly in favour of fat. The storage level in the fat is related to the quantity ingested and varies according to species. Biliary excretion starts shortly after absorption, mainly in the form of hydrophilic metabolites. A part of the excretion products is reabsorbed and again

---

*Quite recently, when this manuscript is being sent to press, Tunstall Laboratory reports additional findings, which modify the above statements. Later studies have shown in both control and test CFI-mice a small number of tumours with some signs of malignancy. This work will be published upon completion.

transported to the liver. Thus, an enterohepatic circulation occurs. About 90% of the total dose is excreted as hydrophilic metabolites in the faeces and about 10% in the urine (Heath *et al.* 1964, Ludwig *et al.* 1964).

Cole *et al.* (1968) dosed male rats, with and without bile fistulae with 0.25 mg/kg dieldrin -C^{14}. The urine and faeces were collected daily. The bile was collected 1, 3, 6, 12 and 24 hours after injection and subsequently at daily intervals. After 5-7 days the animals were sacrificed. The radioactivity in all excreta, bile and tissues was measured by liquid scintillation counting. Over 90% of the excreted activity was found in the faeces from the intact animals or in the bile of the animals with a bile fistula. 50% of the dieldrin administered was excreted within 3 days; 32% had been excreted in the bile after 6 hours.

The enzymes responsible for the epoxidation of aldrin to dieldrin are localized in the smooth endoplasmatic reticulum. Epoxidation of aldrin, heptachlor and isodrin is more rapid in the male rat than in the female (Wong *et al.* 1965). The epoxidation of aldrin to dieldrin occurs in the liver microsomes. Therefore dieldrin may be regarded as a metabolite of aldrin.

In a study with C^{14}-labelled material, it was demonstrated that microorganisms, mosquito-larvae, mammalian liver homogenates, perfused livers and intact mammals, metabolize both aldrin and dieldrin. Rats and rabbits excrete mainly hydrophilic metabolites in faeces and urine. From the urine of C^{14}-dieldrin-fed rabbits six different metabolites were isolated. Of the total urinary excretion, 85% consists of one of the two enantiomorphic isomers of trans-6,7-dihydroxy-dihydro-aldrin (transdiol). This compound has an acute oral LD_{50} of 1250 mg/kg to mice. Of two other metabolites, accounting for 2-4% of the total urinary radioactivity, the structures have been elucidated with a large degree of confidence (Ludwig *et al.* 1966, Korte 1967).

In the urine of the rat a metabolite occurs which is a ketone. The trans-diol found in rabbits is not present in rat urine. The rat, however, excretes dieldrin mainly via the faecal route in the form of a monohydroxy-substitution product of HEOD, the structure of which has now been elucidated (Richardson *et al.* 1968).

The presence of identical hydrophilic metabolites in the urine of rats, after administration of aldrin or dieldrin has been demonstrated by Datta *et al.* (1965), in a study on rats fed diets containing 25 ppm of aldrin and dieldrin for 120 days. Two metabolites were found in the urine. The male rat excreted a much larger quantity of one metabolite than the female rat (in a ratio of 5 : 1). Of the other, the quantities excreted were equal in both sexes. Further work with regard to the structure of these metabolites has shown that they are almost identical, differing only in minor spectrometric characteristics. These differences are probably due to alternative exo- or endo-orientations of the epoxide ring (Damico *et al.* 1968).

4.6 PHARMACODYNAMICS

Moss and Hathway (1964) demonstrated that the solubility of dieldrin in rabbit serum is 4000 times greater than its solubility in water. Dieldrin is primarily located in the erythrocyte contents and in the blood plasma of rabbit and rat, but not in the leucocytes, the platelets and the erythrocyte-stroma. The distribution between plasma and red cells is roughly 2 : 1, according to Weikel et al. (1958). In the red cells, dieldrin is largely associated with haemoglobin and an unknown constituent, also with albumin, α_1- and α_2-globulin and another unidentified component in rabbit serum (Moss and Hathway 1964). The erythrocyte membrane is freely permeable to dieldrin.

Earlier studies with C^{14}- labelled dieldrin had already shown that in the pregnant rabbit dieldrin freely passes the placenta in both directions and is excreted in milk (Bäckström et al. 1965, Hathway et al. 1967). A recent study on dairy heifers fed dieldrin at 0.11 mg/kg body weight during 60 days has confirmed this placental transfer (Braund et al. 1968).

Heath et al. (1964) demonstrated that dieldrin can pass the blood-brain barrier freely in tests with Cl^{36}-dieldrin in rats.

Other studies with rats had shown that, at a given and constant rate of ingestion of tolerated doses of dieldrin, the concentration in the tissues approaches but does not exceed a constant value. Whilst in theory this concentration is never actually attained, in practice constant tissue concentrations are observed if the intake is continued for a sufficient time. In effect a plateau is reached, the concentrations being directly related to the dosage rate. In rats a marked sex-difference was found in this plateau level since at similar dose-levels excretion was less in the female rat than in the male. The half-life of dieldrin in the male rat was found to be 10-11 days and in the female approximately 100 days (Wong et al. 1965, Robinson et al. 1969).

In the two-years' feeding study on rats by Walker et al. (1968), the existence of a plateau-level corresponding with each dose level was confirmed. This plateau was reached in the various tissues of rats on the highest dose level in the course of 6-12 months.

Ibrahim (1964) treated rats at a dietary level of 50 ppm dieldrin for 7 months and a dieldrin blood-level of 0.25 μg/ml was found, without convulsions occurring.

Deichmann et al. (1968) fed female rats on a diet containing 50 ppm dieldrin and sacrificed animals at regular intervals to determine dieldrin levels in blood, liver and adipose tissue. Levels in blood and liver were constant from the 9th day onward, in adipose tissue from the 16th day onward. After this, there was a remarkable constant relationship between dieldrin levels in blood, liver and fat, roughly in the ratio of 1 : 30 : 500.

A chronic feeding study in pigeons by Robinson et al. (1967) indicated that a plateau-level in this species is highly probable.

Walker et al. (1968) describe a two-year study on dogs, fed daily dosages dieldrin of 0.005 and 0.05 mg/kg body weight. A plateau was reached after 12-18 weeks at the lower feeding level and after 18-30 weeks at the higher feeding level. In contrast to the rats, no sex difference was observed in the dog.

Brown et al. (1964) fed 4 beagles at daily oral intakes of 0.4-0.8 mg/kg body weight. Dieldrin concentrations in the blood were determined during 8 episodes of intoxication: these ranged from 0.27-1.27 μg/ml. 2 dogs on 0.2 mg/kg/day reached dieldrin concentrations of 0.11-0.22 μg/ml without signs of intoxication. This suggests a threshold level for dieldrin intoxication > 0.2 μg/ml in the blood for this animal species.

This threshold concentration is applicable to dieldrin regardless of the type or duration of exposure (Keane et al. 1969-b).

There is a linear relationship between log-intake and log-storage for aldrin and dieldrin similar to that occurring for the fat-soluble nutrient vitamin E (Quaife *et al.* 1967).

The storage relationship and the level of the plateau reached varies with the tissues examined: adipose tissue ≫ liver > brain > blood (Robinson *et al.* 1969).

The results of all these studies are consistent with the compartment model hypothesis (Robinson and Roberts 1967): "The correlations are considered to be the result of a functional dependence between the concentrations of insecticide in various tissues and that in the circulating blood or plasma". These relationships are summarized by Robinson and Roberts (1967) as follows:

– "the concentration of a particular organochlorine insecticide in a given tissue is related to that in the other tissues;

– in chronic oral ingestion the concentrations of these compounds in the various tissues are a function of their concentration in the diet;

– chronic oral ingestion of these compounds does not result in a continuous rectilinear increase in their concentrations in the body tissues; where sufficient results are available an asymptotic relationship is found between the time of exposure and the tissue levels;

– on terminating the exposure to these compounds their concentrations in the body-tissues decrease in an exponential manner".

Stoewsand *et al.* (1968) showed that rats on a low protein diet were less resistant to the toxic action of 150 ppm dieldrin in their daily diet than rats raised on a high protein diet. In the latter group there was an increased urinary and faecal excretion of metabolites of dieldrin-C^{14} and a lower storage of dieldrin in body tissues as compared with the other group.

According to Keane *et al.* (1969) in dogs, the time interval before symptoms of poisoning appear, following prolonged daily intake of a relatively constant toxic amount of dieldrin, is directly related to obesity in the sense that obese animals will show signs of intoxication significantly later than lean animals. This was found to be true for all routes of absorption: ingestion, inhalation or percutaneous.

Keane and Zavon (1969) fed mongrel dogs at dietary levels of 0.2 mg/kg/day dieldrin in corn oil 5 days per week for 60 days and observed them for 19 days afterwards. They confirmed the establishment of a storage equilibrium and also the relationship between dieldrin concentrations in blood and in adipose tissue. However, they found that in order to relate the total body-burden to blood or fat concentrations the degree of obesity has to be taken into account.

Thus there is only an indirect relationship between total body-burden and blood or fat concentrations of the insecticide. The greater the total body fat, the greater the total body-burden of dieldrin at a given daily intake (Keane and Zavon 1969).

4.7   CENTRAL NERVOUS SYSTEM AND PERIPHERAL MOTOR EFFECTS

The exact mode of action of aldrin and dieldrin on the central nervous system is still not fully understood. Hathway *et al.* (1965) in summarizing their rat experiments wrote: "The action of dieldrin leads to liberation of ammonia in the brain before the onset of convulsions and throughout their course. The α-ketoglutarate, glutamate, pyruvate and alanine are utilized in an ammonia-binding mechanism. Later in the seizure pattern, this mechanism is inadequate and free ammonia builds up in cerebral tissues. Dieldrin therefore seems to inhibit glutamine synthesis in the brain".

Khairy (1960) studied the effect of chronic dieldrin exposure on the muscular efficiency of rats exposed to diets containing 25 and 50 ppm dieldrin. The time taken to perform a certain exercise was used as the criterion for the muscular efficiency of the rat. He observed a progressive deterioration of muscular efficiency related to the amount of dieldrin administered.

London and Pallade (1964) in long-term experiments with rats fed at dietary levels of 3 mg/kg/day aldrin, 6 days a week for 6 months, followed by 4.5 mg/kg/day for a further 7 months measured chronaxie by applying an electric current to the tail of the rat and determining the pulse duration at varying current strengths which elicited a withdrawal response. They found that a longer pulse duration was necessary in the case of the aldrin-exposed rats when compared with control animals. The same authors studied the effect of acute intoxication with an oral dose of 97 mg/kg aldrin on the same reflex in the rat, testing them 2, 4 and 6 hours after oral dosing. In this instance they found signs of increased excitability.

Ibrahim (1964) examined rats fed dieldrin over a period of 7 months at a dietary level of 50 ppm or treated with a single acute dose of 70 mg/kg body weight dieldrin. When treated with curare, the frequency-tension curves obtained on direct stimulation of the muscle were similarly elevated at the lower frequencies of stimulation. This suggests that the effect of dieldrin is exerted in the muscle itself. Ibrahim also found that the gastrocnemius muscle of dieldrin-treated rats failed to maintain a tetanus compared with those of control rats. She concluded that the increased excitability of the nerve and an increase in the duration of the active state of the muscle would account for the poor maintenance of tetanus, and this could be interpreted as a rapid onset of fatigue, which would be compatible with Khairy's observations mentioned above.

Natoff et al. (1967) in an extension to the above mentioned studies of Khairy, London et al. and Ibrahim, determined rheobase, chronaxie and tetanic frequency values in phrenic nerve-diaphragm preparations of rats exposed to dieldrin at a dietary level of 100 ppm for 26 weeks. The state of incipient intoxication occurring in these rats did not significantly affect rheobase, chronaxie and tetanic frequency values. The same authors, in the same study, using groups of mice fed for 10 weeks on diets containing respectively 0, 5, 10 and 20 ppm dieldrin, found that the exposures did not affect the convulsion threshold of either leptazol or strychnine, whereas acute intoxication with dieldrin selectively and significantly increased the sensitivity to leptazol. The authors conclude: "that acute intoxication with HEOD produces a facilitation of nervous transmission either at the site of excitation of leptazol (cerebral structures) or on the polysynaptic nervous pathway between the sites of action of leptazol and of strychnine (spinal locus)".

4.8   ENZYME INDUCTION

It was previously mentioned in chapter 2.1.5 that many compounds, e.g. ethanol, pyrethrum, butylated hydroxytoluene, phenobarbital and chlorinated hydrocarbon insecticides such as DDT and the cyclodienes chlordane, dieldrin etc., have a stimulatory effect on the drug-metabolizing enzyme system of the liver and are frequently associated with liver enlargement (Burns *et al.* 1963, Fouts 1963, Hart *et al.* 1963, Hart *et al.* 1963-b, Hart *et al.* 1965, Gilbert *et al.* 1967, Conney 1967, Remmer 1967, Kupfer 1967, Rubin *et al.* 1968, Kimbrough *et al.* 1968, Conney 1969).

Other effects have been demonstrated, for example:

– that DDT above a certain minimum level of intake increases the activity of the epoxidase enzymes of the rat liver, an increase which is proportional to the level of intake and is related to age, sex and duration of exposure (Gillett *et al.* 1966);

– that dieldrin and heptachlorepoxide concentrations in the fat of rats, sheep and swine, simultaneously administered dietary levels of DDT of 5 ppm or more, were lower as compared with the concentrations of those animals dosed only with dieldrin or heptachlorepoxide (Street 1964, Street *et al.* 1966);

– that suitable doses of phenobarbital decrease the dieldrin storage level in the fat of rats, but have little effect on the level of dieldrin in the liver. This effect is explained by the authors (Cueto *et al.* 1967) on the basis of increased liver microsomal activity;

– that a single dose of 1 mg/kg body weight of aldrin, dieldrin or chlordane administered 4 days previously, decreases the toxicity of parathion, paraoxon and several other organophosphates to the rat, although there was increased toxicity during the first few hours after pretreatment (Triolo *et al.* 1966, Triolo *et al.* 1966-b, Bass *et al.* 1968).

Street and Chadwick (1967) found that the excretion of polar dieldrin-^{14}C metabolites in faeces and urine in DDT treated female rats, greatly exceeded the excretion of those same metabolites in rats given only dieldrin-[14]C. Clearly DDT stimulates metabolism of dieldrin.

In rat studies Street *et al.* (1966-a) showed that drugs like heptobarbital, aminopyrine, tolbutamine and phenylbutazone were effective in reducing tissue levels of dieldrin and to about the same degree as the reduction produced by DDT.

Peakall (1967) fed male and female White King pigeons 2 ppm dieldrin in their diet for a week. The total amount of dieldrin ingested was about 1 mg per animal. After sacrificing the animals an increased rate of metabolism of steroids by induction of hepatic enzymes was found. Similar results were obtained with 10 ppm dietary DDT, or a total intake of approximately 5 mg DDT per bird.

Liver cells of rats fed 200 ppm of dieldrin in the diet showed effects quite similar to those observed in rats fed phenobarbital. The effects were observed 24 hours after the initiation of the feeding regime and became more marked with increasing time of exposure. These effects were reversible. The electronmicroscopic changes were correlated with an increased activity of certain enzymes, such as those capable of hydroxylating aniline or catalizing the 0-dealkylation of chlorfenvinphos. The activity of other enzymes such as acid phosphatase or glucose-6-phosphatase showed no change (Donninger *et al.* 1967).

That the liver enlargement is associated with a rise in serum alkaline phosphatase, was observed in the top dose group in the recent two-year feeding study on dogs (Walker *et al.* 1968). In this study, the electronmicroscopic changes induced by phenobarbital and dieldrin were similar to those found in rats (Wright *et al.* 1967) and were also reversible. In the dog, however, in contrast to the rat, an increased activity of liver and serum alkaline phosphatase levels was observed at the highest dose-level.

In the chronic toxicity study by Zavon *et al.* (1967) on rhesus monkeys, discussed in 4.3.3, liver cell changes, without detectable enzyme changes were found only in animals receiving dietary dieldrin levels of 1.0 and 2.5 to 5.0 ppm and not in those at lower dietary levels.

Deichmann *et al.* (1968, 1969) fed by capsule, aldrin (0.6 mg/kg/day), DDT (24 mg/kg/day) and a combination of aldrin and DDT (0.3 mg/kg/day and 12 mg/kg/day respectively) per capsule to three groups of 6 adult male beagles, 5 days a week for 10 months. A 4th group served as a control. After this exposure period the dogs were observed for another 12 months. At regular intervals throughout the whole period liver and adipose tissue biopsies were carried out and blood samples taken. Both dieldrin and DDT levels were determined in all samples derived in this way. Dieldrin and DDT levels were higher and took longer to return to normal in those dogs which were fed the combination of half doses of DDT and aldrin. Thus at these relatively high levels of intake of aldrin and DDT there were no signs of these insecticides stimulating each others metabolism, on the contrary there was a proportionately decreased elimination of both compounds from the body.

Street *et al.* (1969) found, after two weeks oral administration to rats, a dietary threshold level for hepatic enzyme induction by cyclodiene epoxides of 1 ppm, as compared with 5 ppm for DDT. According to these authors there is a rather wide species variability for this effect and the rat may be a poor representative species, because of its exceptional sensitivity to hepatic enzyme induction.

In contrast to this enhanced elimination from the body, a combined administration of near-threshold doses of structurally similar compounds, such as the cyclodiene epoxides, may give rise to competitive substrate inhibition, and thus to decreased elimination from the body (Street *et al.* 1969). This leads these authors to conclude: "Thus a bewildering array of species variability is evident which is probably added to by factors of age, sex, level and duration of exposure to the pesticides".

4.9 DISCUSSION

In long-term feeding studies with laboratory animals the first measurable effect in all species is an effect on the liver. This effect, consisting of a stimulation of the drug metabolizing enzyme system of the liver, is associated with an increase in liver weight and, at a later stage, with an increase of the activity of various hepatic and serum enzymes produced by the microsomal system.

This effect was shown to be of a physiological type and to be reversible and similar to the effect caused by various other compounds and drugs such as phenobarbital. This so-called "chlorinated hydrocarbon liver" is a hyperactive and not a damaged liver.

Although this is a measurable effect, it is a sign of exposure and adaptation rather than of injury or intoxication.

This effect is dose-related and reversible, with a dietary threshold level of 1 ppm dieldrin corresponding to 0.05 mg/kg body weight per day in the rat, dog and monkey (Zavon *et al.* 1967, Wright *et al.* 1967, 1968, Walker *et al.* 1966, 1967, 1968, Street *et al.* 1969).

Neither in the dog nor in the rat is the administration of aldrin or dieldrin associated with an increased incidence of malignant tumours.

Only in inbred mouse strains has the administration of dieldrin produced liver nodules, which are similar to those induced by phenobarbitone, have no invasive growth, do not metastasise, cannot be transplanted to other mice and do not have autonomous growth, but on the contrary tend to decrease in size when dieldrin, or indeed phenobarbitone, administration is discontinued. These liver nodules are not being considered as malignant tumours. Further studies are in progress.

4.10 SUMMARY

This survey on the toxicology of aldrin and dieldrin is restricted to the newer literature on these topics. Aldrin and dieldrin have been considered together, because aldrin is readily epoxidized to dieldrin in the body.

Symptoms of intoxication with these compounds are restricted to effects of stimulation of the central nervous system, ultimately culminating in epileptiform convulsions. Death may occur after 6-8 convulsions. Animals recovering from intoxication show no after-effects, neither clinically nor at autopsy.

Underfed animals, with smaller fat deposits, are more susceptible to toxic doses of aldrin and dieldrin both in acute and long-term feeding experiments.

The exact mechanism by which dieldrin acts on the central nervous system is still not fully understood. The severity of symptoms is related to the dieldrin level in the blood. The threshold level for convulsive intoxications in the rat and the dog appears to be 0.25 μg/ml dieldrin in the blood.

The acute oral LD_{50} of these compounds is in the range of 18.8-109.0 mg/kg body weight, depending on animal species and vehicle used, most results being near 60 mg/kg body weight.

After ingestion aldrin and dieldrin are readily and almost completely absorbed from the gastrointestinal tract and transported from the liver to the body. The erythrocyte membrane, the placenta and the blood-brain-barrier are freely permeable to dieldrin. Metabolism takes place in the endoplasmatic reticulum of the liver, where aldrin is readily transformed to dieldrin. At a slower rate dieldrin is metabolized to hydrophilic metabolites, which are then excreted via the bile and the urine. In the rabbit the most important metabolite excreted in the urine is a transdiol of low oral toxicity; in the rat faeces and urine other hydrophilic metabolites have been demonstrated.

The structure of the main metabolites in the urine of rats and rabbits and in the faeces of rats has been established.

In rats about 90% of the metabolites are excreted in the faeces and 10% in the urine. 50% of dieldrin is excreted within 3 days in the male rat; in the female rat this takes longer. This sex difference is not found in the dog.

In all species examined the existence of a steady state of storage at a certain dose level of dietary intake for aldrin and dieldrin has been proved (rat, dog and monkey) or indicated (pigeon). A linear relationship between log-intake and log-storage of aldrin and dieldrin was demonstrated. The storage level is typical and interrelated for various tissues, consistent with the compartment model hypothesis.

On terminating exposure to these compounds their concentrations in the body tissues decrease in an exponential way. The total body-burden is indirectly related to the storage level, and the total amount of body fat needs to be taken into account.

In long-term exposure the initial response to aldrin and dieldrin is an increase in liver weight, followed by indications of hepatic enzyme induction. The threshold level for hepatic enzyme induction in rat, dog and monkey for aldrin and dieldrin is the 1 ppm dietary level. At this 1 ppm dietary level — corresponding to a daily intake of 0.05 mg/kg body weight — however, liver weight is increased in rats and dogs.

Data available indicate that aldrin and dieldrin do not produce malignant tumours in mouse, rat and dog at all tolerated dose-levels in long-term feeding studies.

Chapter 5

General toxicology of aldrin and dieldrin — studies on man

5.1   INTOXICATIONS IN MAN

5.1.1 Symptomatology

When a toxic dose of aldrin or dieldrin has been ingested or has contaminated the skin, a more or less typical syndrome appears from twenty minutes to twenty four hours afterwards, *viz.* headache, dizziness, nausea, general malaise, vomiting, followed later by muscle twitchings, myoclonic jerks and even convulsions. Death may result from anoxemia (Nelson 1953, Princi 1957, Hayes 1957, 1963, Hoogendam *et al.* 1962, 1965, Kazantzis *et al.* 1964, Schafer 1968).

The duration of the interval between oral intake or skin contact and onset of symptoms as well as the clinical picture depend on the dose absorbed, with massive over-exposure it is possible that convulsions occur without prodromi.

There is no fever or change in bloodcount or in bloodchemistry. However, abnormal EEG patterns showing spike and wave complexes and multiple spike and wave discharges, or in less serious intoxications, bilateral synchronous thêta discharges may be seen. The diagnosis must be confirmed by determination of the insecticide concentration in the blood.

Hodge *et al.* (1967) estimate the lethal dose of aldrin and/or dieldrin for an adult man to be about 5 gram.

In workers who handle aldrin and dieldrin a slow build-up of the insecticide in the body occurs if the daily intake exceeds the daily excretion. If toxic levels are reached the following prodromal signs may appear: headaches, lassitude, fatigue, loss of appetite, weight loss, insomnia, frequent nightmares, inability to concentrate, loss of memory, hyperirritability, hyperexcitability, paresthesia, myoclonias and black-outs.

5.1.2 Cases in medical literature

Spiotta (1951) was the first to describe a case of acute convulsive aldrin poisoning in a 23 year old man who ingested an aldrin emulsifiable concentrate (e.c.) in an attempted suicide. This resulted in convulsions accompanied by typical EEG changes. The man fully recovered.

Nelson (1953) described 3 cases of convulsive intoxication in workers formulating a 25% aldrin dust-concentrate as a result of ignoring safety directions.

Hayes (1957, 1959) reported nearly 100 convulsive intoxications in spray operatives handling dieldrin in malaria-eradication campaigns in 5 tropical countries.

Bell (1960) described a convulsive attack due to gross over-exposure in an aldrin packer in Australia.

Kazantzis *et al.* (1964) described 3 convulsive intoxications in men formulating aldrin. All 3 fully recovered.

Hoogendam *et al.* (1962, 1965) discussed 17 convulsive intoxications from our aldrin, dieldrin and endrin manufacturing and formulating plants. All workers made a full recovery.

Van Raalte (1965) collected from the world literature all cases of fatal poisoning by aldrin and dieldrin reported at that time, 13 in total: 4 suicides, 3 due to accidental ingestion, 5 due to accidental contamination and only one, a spray operative, in the course of occupational exposure. To the best of our knowledge no cases of fatal poisoning have occurred in the course of aldrin and dieldrin manufacture and formulation.

5.1.3 Treatment and prognosis of intoxication

First aid in the case of accidental skin contamination should consist of immediate removal of all contaminated clothes, including underwear and washing of the contaminated skin and hair with soap and water. All contaminated clothing should be changed and laundered before re-use.

In the case of ingestion vomiting should be induced and the stomach emptied as quickly as possible.

If the patient is unconscious a free airway should be ensured. If respiration has stopped artificial respiration should be employed.

Medical treatment is largely symptomatic and supportive and directed against convulsions and anoxemia. Carbo absorbens may be given. Sodium sulphate may be administered as a laxative. Oily laxatives or milk should not be given. Morphine, epinephrine and ncradrenaline are contra-indicated.

Administration of phenobarbitone may both control prodromal symptoms and prevent convulsions. When convulsions do occur barbiturates should be given slowly intravenously, *e.g.* thiopentone sodium 10 mg/kg with a maximum of 750 mg for an adult. The administered dose should be sufficient to control convulsions.

Following acute intoxication, oral administration of phenobarbitone at 2-5 mg/kg/day (150-300 mg/day for an adult) for a period of a few days to up to two weeks, is recommended, the dose and the length of the therapy depending on the subjective well-being of the patient and the insecticide concentrations in the blood.

An unobstructed airway must be maintained. When needed, oxygen and/or artificial respiration must be given.

If prompt and adequate treatment is given, then death can be prevented and even the severest intoxication will recover completely within some weeks (Princi 1957).

Furthermore EEG's will return to a normal pattern within months (Princi 1957, Hoogendam *et al.* 1962, 1965).

5.2   PHARMACODYNAMICS

A pharmacodynamic study was carried out in Tunstall Laboratory with four groups of volunteers (3 or 4 subjects per group), who were given daily measured doses of HEOD in gelatin capsules for two years   (Hunter *et al.* 1967, 1969). The dosage regimes were resp. 0, 10, 50, and 211 μg HEOD per day which, added to the dietary intake, increased the daily dose to approximately 14, 24, 64 and 225 μg/day. The dieldrin concentrations in the blood and adipose tissue were determined, urinalysis, EEG studies, polygraphic recording of cardio-respiratory function, electromyographic studies were performed and blood chemistry (estimation of blood plasma protein and urea, activity of plasma alkaline phosphatase, SGPT, SGOT and the activity of erythrocyte and plasma cholinesterases) was studied. Full clinical examinations were made after 3, 9, 15, 18 and 24 months of exposure. The subjects selected were 13 healthy men without a history of recent occupational exposure to pesticides. Their age ranged from 21 to 52 years.

The results of all these observations showed that the only changes observed in the exposed subjects were increases in the concentration of HEOD in adipose tissue and blood. A relationship was established between concentrations of HEOD in whole blood, the adipose tissue, the daily dose of HEOD and the duration of the exposure. This relationship is of a curvilinear type with a finite upper limit, an asymptote. Thus by this study three important facts were established:

1. no anomalies
2. equilibrium at plateau reached after 9-12 months
3. mathematical relationship between HEOD concentrations in blood and fat and daily intake at equilibrium.

The upper levels of concentration were 0.0202 μg/ml in the blood and 2.85 μg/g in adipose tissue of the men given 211 μg/day. As mentioned above the total daily exposure of these men was 211 μg + an estimated intake from dietary sources of 14 μg/day, *i.e.* 225 μg/day.

The concentration of HEOD in adipose tissue at the state of equilibrium is related to that in whole blood, the ratio of these concentrations is approximately 136 (confidence limits p=0.95 : 109 - 170).

The relationship between the concentration of HEOD in these two tissues and the total daily intake at the steady state of storage is:

$$\text{Dietary intake of HEOD (in } \mu g/man/day) = \frac{\text{Conc. of HEOD in blood } (\mu g/ml)}{0.000086}, \text{ or}$$

$$= \frac{\text{Conc. of HEOD in adipose tissue } (\mu g/g)}{0.0185}$$

An estimate of the half-life of HEOD in the blood of man, based on the decrease in tissue levels during the first 9 months of the post-exposure period is approximately 369 days. These relationships are similar to those found in rats and dogs (Walker *et al.* 1968).

There is a linear relationship between log-intake and log-storage (Hunter *et al.* 1967).

Of the 12 adult human volunteers with known oral exposures, body weights and total body fats, as measured by skinfold thickness, varied little over the 2 year test period. According to Hunter *et al.* (1968): "The highest concentrations of HEOD in adipose tissue were found in the leanest subjects and these subjects also exhibited the smallest body-burden. On the other hand, the proportion of the total exposure dose retained in adipose tissue was highest in those subjects with the greatest total body fat". Keane *et al.* (1969) found a similar relationship between body-burden and concentration in adipose tissue in dogs.

The question has arisen that severe weight loss might possibly cause the release of dieldrin from the storage depots, increasing the concentration in the blood and thus precipitating manifestations of intoxication. In order to investigate this question, observations were made on subjects following severe reducing diets for slimming purposes and on patients undergoing elective surgery (Hunter *et al.* 1968). Blood levels of dieldrin were determined before and during the period of slimming and in the latter group pre- and post-operatively and during the operation. Neither significant increases in blood concentrations nor any symptoms of intoxication were observed.

Van Raalte (1965) calculated the risks arising from extreme weight loss: "In man it is well known that under no circumstances more than 500 gram of fat can be lost per day and this rate of loss of fat would not be sustained for a long period. At a fat level of 0.2 ppm 500 gram of fat would release 0.1 mg of dieldrin over a period of 24 hours, or in a 70 kg man 0.0014 mg/kg/day, which is less than one ten thousandth of the intravenous LD_{50} for rats, *i.e.* | 15.2 mg/kg/day". Observations of industrial workers show that prolonged total daily intake of considerably more than 0.1 mg is not associated with intoxication. "It is therefore unlikely that the maximum amount of weight loss in man could give rise to any intoxication due to the release of dieldrin stored in fat".

5.3 ENZYME INDUCTION

In the pharmacodynamic study of Hunter *et al.* (1967, 1969), pp'DDE concentrations in adipose tissue of the volunteers were determined at the same time as the dieldrin concentrations. In this respect Robinson (1969) stated: "It appears that compounds which induce changes in the activity of hydroxylating enzymes in the smooth endoplasmatic reticulum of cells will, consequently, increase the rate of metabolism of the organochlorine insecticides. If the rate of transformation of these compounds is increased then the concentration in tissues will decrease. A preliminary examination of the concentrations of pp'DDE in the blood of the volunteers given known daily amounts of 211 μg HEOD for two years has not shown any significant decrease in the concentration of pp'DDE relative to that in the control group, and it is tentatively concluded that total intakes of up to 225 μg per day of HEOD do not increase the rate of metabolism of pp'DDE in man".

In the light of a dietary threshold level for cyclodiene epoxides of 1 ppm for hepatic enzyme induction in the rat and considering that this species is very sensitive to this response, Street observed: "The reported quantities of these organochlorine insecticides present in diets of U.S. residents (0.0014 mg/kg body weight daily or about 0.024 ppm in the diet, based on food consumption of the 16 to 19 year old male) seem to be well below the threshold dosages for significant enzyme induction" (Street *et al.* 1969).

5.4 EXPOSURE OF THE GENERAL POPULATION

5.4.1 Determination of dietary intake

More than 90% of the total intake of the general population of organochlorine insecticides in Western Europe and the U.S.A. result from residues in food (see chapter 1.2 and 1.3). These dietary intakes have been monitored for many years in various countries by means of market-basket surveys and studies of complete prepared meals (chapter 1.5.1).

Total diet studies in the U.S.A. have been reported at regular intervals since 1961 (Mills 1963, Mills *et al.* 1964, Williams 1964, Cummings 1965, Duggan *et al.* 1966, Duggan *et al.* 1967-a, Duggan *et al.* 1967-b). Duggan reviewed the data obtained between June 1964 and April 1966 from total diet samples collected from 25 cities across the U.S.A. (Duggan *et al.* 1967-b) and estimated the combined average total intake of all the organochlorine pesticides in the general population throughout this period to be approximately 0.1 mg/man/day. This figure was based on the high food-intake of a 19-year old boy, who would ingest a daily quantity of 4.0 kg of food and drink (Duggan *et al.* 1967-a). This quantity is approximately double the average quantity and therefore these figures should really be halved if they are compared with those from other countries. Of the total intake 75% was accounted for by DDT and its metabolites, and approximately 33% as DDT itself.

The total diet data reported for the U.S.A. in "The regulations of pesticides in the U.S.A. 1967" for the years 1965-1967 are summarized in Table 4.

Table 4

DIETARY INTAKE OF ORGANOCHLORINE PESTICIDES IN THE U.S.A.

	Intake in μg/kg/day in the U.S.A.				*Average intake over 3 years in μg/man/day*
	1965	*1966*	*1967*	*average*	
Total organochlorine pesticides	1.2	1.6	1.2	1.3	91
DDT + analogues	0.9	1.0	0.8	0.9	63
HEOD	0.09	0.13	0.06	0.09	6.3

It is understood, that continued monitoring of human fat samples in the Chicago area confirms the low levels found by the Chicago Health Research Foundation from 1962 to 1966 (Hoffman *et al.* 1964, Hoffman *et al.* 1967). Accordingly, Hoffman (1968) considered that this data agreed with the results of recent food analyses in the U.S.A. which showed no increase in insecticide content.

In Great Britain, analysis of complete prepared meals show HEOD intakes of 19.9 μg/man/day in 1965 (Robinson *et al.* 1966), decreasing to 12.6 μg/man/day in 1967 (McGill *et al.* 1969), and to about 0.1 μg/kg body weight or about 7 μg/man/day in 1969 (Report of the "Wilson committee" 1969).

This daily intake of about 0.1 μg/kg body weight in the U.S.A. and the U.K. is similar to the A.D.I. advised by the WHO, and 500 times smaller than the no-toxic-effect level of 1 ppm (= 50 μg/kg body weight) in the rat, dog and monkey.

5.4.2 Dieldrin levels in adipose tissue

As with DDT most information on tissue levels of HEOD relates to adipose tissue, and these data have been summarized in Table 5. Only determinations carried out by gas liquid chromatography (GLC) have been included. Some authors give arithmetic means, while others have calculated the geometric mean. With a skew distribution of data, and this kind of data always has a skew distribution, the geometric mean is the more appropriate measure and gives a somewhat lower figure than the arithmetic mean.

Table 5 shows that the concentration of dieldrin stored in the body fat of the population of the United Kingdom and the U.S.A. has reached a constant level (Hunter *et al.* 1963, Robinson *et al.* 1965, Hayes 1966, Robinson *et al.* 1966, Hoffman 1968). This is in agreement with the total diet data reported from the U.S.A. and summarized in Table 4.

Robinson and Roberts (1968) rightly point out that: "Most surveys contain samples from both men and women and the applicability of the above relationships (the formula of Hunter *et al.* 1969) to women has not been established. However, it is pertinent to point out that the concentrations in the adipose tissue of women tend to be lower than those found in men, although the differences are generally not statistically significant. It is assumed therefore, that no gross errors are involved in using samples that include both sexes".

Edmundson *et al.* (1968) reported on dieldrin concentrations in adipose tissue samples from southern Florida, all from subjects accidentally or violently killed. The dieldrin concentrations did not differ when the sample was stratified as to age, race or sex. Neither did Robinson and Roberts (1968) find evidence that there is a significant difference of HEOD levels in adipose tissue in different ethnic groups in the U.S.A. The situation may be different in the case of DDT, where Davies *et al.* (1968) found significant differences in adipose tissue levels between different ethnic groups.

Radomski *et al.* (1968) (see also Deichmann *et al.* 1968) determined organochlorine levels in liver, brain and other tissues obtained at autopsies of 271 subjects in the U.S.A. They found a significant correlation between the pesticide levels in the brain and in adipose tissue. In studying relationships between organochlorine pesticide concentrations, mainly DDT, and disease, they found consistently high pesticide concentrations in adipose tissues from patients with cirrhosis of the liver, carcinoma or hypertension. The authors concluded that from their data it was impossible to determine whether the disease caused an increased pesticide body-burden or vice versa, or indeed, whether both were effected by a still different cause. However, even these high levels were below those usually found to be associated with intoxication. Therefore the former assumption appears the more likely one.

Casarett *et al.* (1968) examined organochlorine pesticide levels in adipose tissue samples from 44 autopsies in Hawaii. Subjects with the highest total residues in the tissues were those with a combined evidence of emaciation, a variety of carcinoma and extensive focal or generalized pathological conditions of the liver. Those dying a sudden death, such as by myocardial infarction appeared to be grouped around the center of the distribution of organochlorine levels. Weight loss, particularly if sudden, might raise tissue levels, as storage depots are depleted; liver abnormalities on the other hand may be expected to influence the metabolism of organochlorines.

Table 5

HEOD LEVELS IN ADIPOSE TISSUE OF GENERAL POPULATION

Country	Year	Type	No. of samples	HEOD conc. in $\mu g/g$ mean + range	Reference
United Kingdom	1961-62	N	131	0.21** 0.02-1.20	Hunter et al. 1963
,,	1963-64	N+B	66	0.21** 0.02-1.08	Egan et al. 1965
,,	1964	N+B	100	0.23** 0.02-1.08	Robinson et al. 1965
,,	1965	N	101	0.23** trace-1.80	Cassidy et al. 1967
,,	1966	N	44	0.22** 0.10-0.73	Robinson et al. 1966
,,	1966	B	53	0.21** —	Hunter et al. 1967
,,	1965-67	N	248	0.17** n.d.-1.00	Abbott et al. 1968
,,	1967	B	18	0.27** —	Hunter et al. 1967
,,	1968	B	24	0.10** —	Robinson et al. 1968
U.S.A.	1961-62	B+N	28	0.15* 0.02-0.36	Dale et al. 1963
,,	1962-63	N	282	0.11* n.d.-1.00	Hoffman et al. 1964
,,	1964	N	25	0.29* 0.03-1.15	Hayes et al. 1965
,,	1964	N	64	0.31* 0.07-2.82	Zavon et al. 1965
,,	1965	N	42	0.22* n.d.-0.70	Fiserova-Bergerova et al. 1967
,,	1965-67	N	146	0.22* n.d.-0.77	Edmundson et al. 1968
,,	1967	N	221	0.14* 0.01-1.39	Hoffman et al. 1967
,, (Hawaii)	1967-68	N	30	0.063* —	Casarett et al. 1968
Australia	1965	B	53	0.046** 0.01-0.43	Bick 1967
,, (West)	1966	N	12	0.67* 0.25-0.99	Wasserman et al. 1968
Canada	1966	N	35	0.22* 0.07-0.53	Brown 1967
India (civilians)	1964	N+B	24	0.03* n.d.-0.36	Dale et al. 1965
,, (military)	1964	N+B	11	0.06* 0.01-0.19	Dale et al. 1965
Italy	1966	N+B	22	0.45* 0.03-0.99	Del Vecchio et al. 1967
Netherlands	1966	N	11	0.17** 0.05-0.50	de Vlieger et al. 1968
New Zeeland	1966-67	N	52	0.27* n.d.-0.77	Brewerton et al. 1967
Denmark	1965	N	18	0.20* 0.10-0.34	Weihe 1966

* arithmetic mean   n.d. = not detected   N = necropsy
** geometric mean   B = biopsy

"It appears that the concentrations found in necropsy samples should be used with caution in attempting to assess dietary intakes and that, in particular, residues in specimens from terminal hospital patients should not be used for this purpose" (Robinson *et al.* 1968).

Hoffman (1968) in a study of 688 patients who died from a variety of diseases in Chicago hospitals could find no significant correlation between organochlorine levels in the various organs and the presence or absence of abnormalities in the organs or tissues. Similarly there was no significant correlation between the pesticide levels and the presence or absence of cancer. He concluded that there is no evidence available that the pesticides, in the concentrations found chronically stored in the body, have any harmful effect. Robinson *et al.* (1965) arrived at the same conclusions.

Fiserova-Bergerova *et al.* (1967) found in a study of 71 autopsies of accidentally killed men and women in Florida, that organochlorine levels in fat were approximately 10 times those in the liver and approximately 100 times those in kidney, brain and gonads. With the exception of the liver, as the centre of metabolism and excretion, the authors supposed that this ratio may be related to the fat content of these tissues.

De Vlieger *et al.* (1968) collected samples of brain tissue, liver and adipose tissue from 11 routine autopsies in Holland. Geometric means and ranges for the HEOD concentrations in these tissues were:

— white matter of the brain   0.0061 ppm (0.002-0.041)
— grey matter of the brain   0.0047 ppm (0.002-0.018)
— adipose tissue   0.17   ppm (0.05-0.50)
— liver   0.03   ppm (0.007-0.081)

A significant relationship was found to exist between the HEOD concentrations in the various tissues, a relationship of the same type as that found between the concentrations in whole blood and adipose tissue (Hunter *et al.* 1969). This relationship may be ascribed to a dynamic partition process between the circulating blood and the

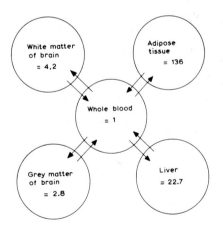

Fig. 8. Distribution of HEOD between blood and tissues in man (modified after De Vlieger *et al.*, 1968).

other tissues. A compartmental model explaining the partitioning of organochlorine insecticides in vertebrates has been proposed by Robinson (1967).

De Vlieger *et al.* (1968) suggested a tentative scheme for the distribution of dieldrin between the various tissues. This scheme is reproduced in Fig. 8, but the figures have been updated by recalculation conforming to the latest empirical formula of Hunter *et al.* (1969).

5.4.3 Dieldrin levels in blood

Compared with the large amount of data available concerning organochlorine levels in adipose tissue, data concerning blood levels of organochlorine insecticides in the general population are relatively scarce. These are summarized in Table 6.

The concentration of HEOD in whole blood samples of healthy members of the general population in the United Kingdom have not changed significantly during the years 1962 up to and including 1968. The current concentrations found in blood plasma and serum of healthy members of the general population of the U.S.A. are of about the same order (Hayes and Curley 1967).

Table 6

HEOD LEVELS IN THE BLOOD OF THE GENERAL POPULATION

Country	Year	Type	No. of samples	HEOD conc. in µg/ml mean+range	Reference
United Kingdom	1962	W	20	0.0025** 0.0005-0.0100	Brown *et al.* 1964
,,	1964	W	44	0.0014** 0.0006-0.0050	Robinson *et al.* 1966
,,	1965	W	25	0.0017** 0.0006-0.0087	Hunter *et al.* 1967
,,	1966	W	55	0.0018** 0.0010-0.0043	Hunter *et al.* 1967
,,	1968	W	18	0.0009**	Robinson *et al.* 1968
U.S.A.	1965	W	10	0.0014* n.d.-0.0028	Dale *et al.* 1966
,,	1966	W	10♂	0.0055* 0.0010-0.0129	Dale *et al.* 1967
,,	1966	W	10♀	0.0018* 0.0006-0.0045	Dale *et al.* 1967
,,	1968	P	10♀	0.0003* 0.0001-0.0012	Curley *et al.* 1969

* arithmetic mean	n.d. = not detected	W = whole blood
** geometric mean		P = blood plasma

5.5 OCCUPATIONALLY EXPOSED WORKERS

Aldrin and dieldrin have been manufactured in the U.S.A. since 1950 and in the Netherlands since 1954. Workers in these manufacturing plants have been medically supervised for many years (Hoogendam *et al.* 1962, 1965, Jager *et al.* 1967, Hayes *et al.* 1968, Hunter *et al.* 1968, Van Dijk 1968).

Medical data from the Dutch manufacturing and formulating plant have been reported by Hoogendam *et al.* in 1962 and 1965. Of 300 workers employed in aldrin, dieldrin and endrin manufacture for periods up to 9 years, including — after recovery — 17 men who had a convulsive intoxication, none showed clinical symptoms or any other anomaly in any parameter examined: anamnesis, general physical examination and routine laboratory tests. Neither were changes found in EEG tracings and liver function tests (thymol turbidity tests, SGOT and SGPT). The pattern and rate of absenteeism was not different from that of other industrial workers. There was no case of permanent, partial or complete incapacity. The only specific changes occurring in cases of intoxication were temporary typical EEG changes as described in 5.1.1.

Van Dijk (1968) evaluated serum alkaline phosphatase levels of 49 plant operators manufacturing aldrin and dieldrin and found that there was no relationship between the serum alkaline phosphatase and the length of exposure to these insecticides or the HEOD level in the blood.

Hayes *et al.* (1968) reported on the concentration of HEOD in blood and fat of workers in the U.S.A. with occupational exposure to aldrin and dieldrin for periods up to 19 years. The mean (arithmetic) concentration of dieldrin in plasma of 28 workers was $0.0247\ \mu g/ml$ (range 0.0012-0.1370) and the corresponding mean concentration in the fat of these workers was $6.12\ \mu g/ml$ (range 0.60-31.96). Hayes considered it probable that these men had reached a state of equilibrium as regards intake and storage and calculated that their equivalent oral intakes were approximately 0.72-1.1 mg/man/day. No significant relationship was found between the concentration of dieldrin in blood and fat and the incidence of sick-leave.

Some concentrations of HEOD in the blood reported in the literature are mentioned below. The major difficulty is that for the estimation of a toxic HEOD level in blood only levels in blood samples taken directly after intoxication occurred can be considered.

In the literature we found two blood levels quoted for samples taken directly after intoxication occurred: $0.53\ \mu g/ml$ HEOD (Kazantzis *et al.* 1964) and $0.32\ \mu g/ml$ HEOD (Robinson *et al.* 1966). Other levels by Kazantzis *et al.*: 0.10, 0.28, 0.13 and $0.13\ \mu g/ml$ HEOD were in samples taken respectively 4, 1, 1, and 1¼ months after the accident occurred. The only conclusion that may be drawn from these data is that HEOD levels at the time of intoxication must have been higher and, in case of acute over-exposures much higher, than at the time of determination. The same reasoning applies to the levels mentioned by Brown *et al.* (1964): 0.20, 0.37 and $0.53\ \mu g/ml$, respectively 2, 3, and 3 weeks after intoxication, and the $0.28\ \mu g/ml$ HEOD 10 hours after intoxication (Dale *et al.* 1966). Some high insecticide levels found in occupationally exposed workers who remained in perfectly normal health were: $0.12\ \mu g/ml$ (Kazantzis *et al.* 1964), $0.26\ \mu g/ml$ (Robinson *et al.* 1966) and $0.19\ \mu g/ml$ (Dale *et al.* 1967).

All of these data support the contention that a threshold level for intoxication of $0.15-0.20\ \mu g/ml$ HEOD in blood, as suggested by Brown *et al.* (1964) which is also used in our practice, is conservative.

5.6   DISCUSSION

In 5.4 three different methods were discussed for the assessment of the aldrin-dieldrin exposure of the general population. Total diet studies in the United Kingdom and the U.S.A. show that the amount of dieldrin daily ingested by the general population of these countries has not increased over the last few years.

Duggan's estimated daily intakes of dieldrin for the years 1965-1967 in the U.S.A. at 6.3 µg/man/day are somewhat lower than the British data of 19.9 µg/man/day in 1965, and 12.6 µg/man/day in 1967, and are in agreement with current data from the U.K. (Report "Wilson Committee" 1969) and we might conclude that daily dieldrin intakes in the U.S.A. and the U.K. are somewhere in the same range of 6.3-12.6 µg/man/day.

The average daily intake of dieldrin by a population can also be calculated from the mean levels of dieldrin in the blood or adipose tissue by the formula developed by Hunter and Robinson (1967, 1969). The daily intake in the U.S.A. calculated on that basis and using adipose tissue level data, ranges from 6 to 17 µg/man/day. For the United Kingdom the mean estimates derived from the concentration of HEOD in fat are in the range of 2.7 to 11.9 µg/man/day (Robinson et al. 1968, McGill et al. 1968).

Similar calculations based on the average HEOD blood levels found in the general population in the U.S.A. and the U.K. give   daily dieldrin intakes of 22 and 19 µg/man/day respectively.

It is clear from the above mentioned calculations that there is a fair agreement between the results of the 3 methods of assessment of exposure of the general population to aldrin and dieldrin.

Basing their calculations on the dieldrin levels in both adipose tissue and blood from the general population the authors agree that these levels are not increasing, and that on the contrary there are indications of a tendency to decrease.

Robinson (1967), Fiserova-Bergerova et al. (1967) and De Vlieger et al. (1968) all conclude that dieldrin is distributed in the body according to a well defined pattern, probably proportional to the lipid content of the various tissues. However, hepatic dieldrin concentrations are proportionally higher, probably as a result of the metabolic and excretory function of this organ. Dieldrin in blood largely represents the migratory phase, which keeps tissue levels in equilibrium.

The storage, or rather the presence, of dieldrin in adipose tissue is only passive as regards any effect on the total organism in the sense that these dieldrin molecules are at any given moment in time not available for action or interaction in pharmacological or toxicological sense (Hunter et al. 1968).

5.7   SUMMARY

Symptomatology of aldrin-dieldrin intoxication as well as its medical treatment are described. Some data on intoxication from the medical literature are summarized. It appeared that all cases of fatal intoxication resulted from acute over-exposure. In all non-fatal cases of intoxication that have been reported full recovery to normal health took place within weeks or a few months at the most.

Pharmacodynamic studies with human volunteers showed that:

1. the ingestion of amounts of dieldrin up to 225 μg/man/day over a period of two years did not have any effect on the health of male volunteers, neither was there any anomaly in the results obtained from the extensive laboratory tests, including the measurements of hepatic and nervous system function;

2. a steady state of storage developed which was associated with the continued intake of specified quantities of dieldrin;

3. a mathematical relationship exists between the level of intake and of storage in blood and adipose tissue, i.e. a linear relationship between log-intake and log-storage per individual.

This latter relationship permits the calculation of levels of intake or absorption by the simple determination of dieldrin levels in blood, or in fat, of people with dieldrin in the steady state of storage, i.e. at the equilibrium between intake and excretion.

These studies also show that when exposure to dieldrin is discontinued tissue levels fall in the same exponential way as that in which build-up took place.

Calculations, as well as actual determinations, show that there is no risk of dieldrin intoxication arising from starvation or excessive slimming-cures by members of the general population.

Arising from the pharmacodynamic studies employing doses of up to 225 μg/man/day HEOD for two years, the measurements of pp'DDE levels in the blood indicate that drug metabolizing enzyme systems are not stimulated.

Dieldrin exposures of members of the general population in the U.S.A. and the U.K., as measured by total diet studies and as calculated from dieldrin levels in adipose tissue as well as in blood, are currently in the range of 2.7 - 22 μg/man/day, with an overall average of 7 μg/man/day, corresponding to 0.1 μg/kg body weight per day. These levels have shown a tendency to decrease over the last few years, and certainly not to increase.

According to the formula of Hunter et al. (1969) this average daily dieldrin intake of 7 μg/man/day for the general population in the U.S.A. and the U.K. corresponds to an average dieldrin level in blood of 0.0006 μg/ml, a very low level when compared with a conservative intoxication threshold level of 0.20 μg/ml in blood. This average daily intake of 0.1 μg/kg body weight is 500 times smaller than the dietary no-toxic-effect level of dieldrin in the rat, dog and monkey, which is 50 μg/kg.

78

Chapter 6

General toxicology of endrin — animal studies

6.1 Introduction
6.2 Acute toxicity
6.3 Subacute and chronic toxicity
6.4 Carcinogenesis
6.5 Metabolism and biochemistry
6.6 Pharmacodynamics
6.7 Discussion
6.8 Summary

6.1 INTRODUCTION

Although endrin is a stereoisomer of dieldrin, its pharmacokinetics differ substantially from the latter compound, as will be discussed later. Endrin (and endosulfan)....
"cannot reasonably be called persistent in vertebrate animal bodies, although we recognise that little information is available on their persistence in soil or the environment generally" (Report by the Advisory Committee, the "Wilson Committee" 1969).

6.2 ACUTE TOXICITY

Endrin is readily absorbed into the body following ingestion, inhalation or via the intact skin.

The symptoms noted during endrin intoxication are similar to those seen in cases of intoxication from aldrin, dieldrin (see chapter 4.2) and Telodrin. Nelson *et al.* (1956) found a hypersensitivity to sound, touch and pressure stimuli in rats fed high dietary endrin levels (50 and 100 ppm a day).

Following the administration of a single oral toxic dose, convulsions with typical EEG changes appeared after a latent period of 45-60 minutes (Speck *et al.* 1958).

A single intravenous toxic dose of 2 mg/kg (Revzin 1966) in pigeons produced an impairment of different central nervous system functions and CNS-directed convulsions. All these endrin-induced toxic CNS effects could be blocked by pentobarbital in sufficient doses.

The acute toxicity of endrin is higher than that of aldrin and dieldrin. The LD_{50} of course varies with the animal species and the vehicle used. A summary of LD_{50} values, selected from the literature is tabulated in Table 7. In some of the tests differences in toxicity were noted between the sexes, *e.g.* in some experiments female rats were more susceptible than male rats. In most species signs of intoxication can be expected with a single dose in the order of 2 mg/kg body weight. In all animals surviving acute intoxication, recovery is complete and without sequelae (Zavon 1961).

Table 7

ACUTE TOXICITY OF ENDRIN IN EXPERIMENTAL ANIMALS

Animal species	Route of administration	Formulation	LD$_{50}$ in mg/kg bodyweight		References
			male	female	
Rat – young	oral	peanut oil	28.8	16.8	Treon et al. 1955
Rat – adult	”	”	43.4	7.3	”
”	”	”	40.0	–	Speck et al. 1958
” – CFE	”	20% e.c.	6.6	3-4	Muir 1968
”	”	2% f.s.d.	2-5	6.6	”
Rabbit	”	peanut oil	7-10	–	Treon et al. 1955
Guinea pig	”	”	36 est.	16 est.	”
Monkey	”	”	3 est.	3 est.	”
”	”	–	12.0	–	Barth 1967
Rat – CFE	dermal	20% e.c.	10.9	–	Muir 1968
”	”	2% f.s.d.	31.5	92	”
Pigeon	intra-venous	–	1.2-2.0	1.2-2.0	Revzin 1966

CFE = Carworth Farm ”E” Strain   e.c. = emulsifiable concentrate
est. = estimated   f.s.d. = field strength dust

6.3 SUBACUTE AND CHRONIC TOXICITY

In the rat

Nelson et al. (1956) fed diets containing 0, 1, 5, 25, 50 and 100 ppm endrin to adult rats, 5 males and 5 females to a group, for 16 weeks. They concluded that: ”The serum alkaline phosphatase values were higher among rats consuming endrin than in the control group. All rats receiving 100 ppm of endrin died within the first 2-week period. Mortality rates indicated that male rats were significantly more susceptible to the toxic effects of endrin at the lower levels (1 and 5 ppm) than the females. There was a loss of weight in all rats ingesting endrin. The greatest weight loss occurred in rats consuming the two highest levels of endrin. The total average feed consumption of endrin-fed rats was less than that of the control group”.

In a two-year experiment, groups of rats of each sex were given diets containing 1, 5, 25, 50 and 100 ppm of endrin. Concentrations of 50 and 100 ppm were lethal within a few weeks. The concentration of 25 ppm increased the mortality rate of the females. Non-survivors at the three highest levels exhibited diffuse degeneration of the brain, liver, kidneys and adrenal glands. The survivors in the two higher levels showed degenerative changes only in the liver, while those fed at lower levels did not show any organ pathology. The level of 5 ppm caused an increase in liver/body weight ratio in males and an increase in kidney/body weight ratio in females. One ppm was the no-effect level in this two year rat study (Treon et al. 1955).

In the dog

In a series of experiments, dogs, 1 or 2 of each sex to each group, were fed diets containing from 1 to 50 ppm (0, 1, 2, 3, 4, 5, 8, 10, 20, 25, 50 ppm) endrin for 18 months. Two out of 4 fed a diet containing 8 ppm and 1 dog on 5 ppm died; at higher levels all dogs died. The two surviving dogs on 8 ppm were kept on the diet for about 6 months and then sacrificed. Increased organ/body weight ratios for the liver, kidney and brain were found and histopathological examination showed slight degeneration of kidney tissue. Three dogs on 4 ppm of endrin survived (Treon *et al.* 1955).

In an experiment of approximately 19 months duration groups of 4 beagles (2 males and 2 females) were placed on diets containing 1 and 3 ppm of endrin. All dogs on 3 ppm had increased organ/body weight ratios for the kidney and the heart. In some of the female dogs on 1 and 3 ppm a renal abnormality, characterized by slight tubular vacuolisation similar to that present in the female control dog and not dose-related in its severity, was observed (Treon *et al.* 1955). Thus the no-effect level in this study was 1 ppm.

In cattle and sheep

Cattle and sheep were not affected by 5 ppm of endrin in their diet for 112 days (Radeleff 1956).

In fowl

Seven day-old chickens were unaffected by a ration containing endrin at levels of 1.5 and 3 ppm for 42 days. When the concentration of endrin was either 6 or 12 ppm the chicks became highly excitable, failed to gain as much weight as the controls and the survival rates for the 12 weeks were 85% and 5% respectively (controls 100%) (Sherman and Rosenberg 1954).

6.4   CARCINOGENESIS

Only a few carcinogenicity studies with endrin have been published. To cite the results of Treon (1956): "Six groups of rats, each consisting of 20 male and 20 female rats were fed over the period of two years on diets containing endrin to the extent of 100, 50, 25, 5, 1 and 0 ppm respectively. Pathological examinations were made of the tissues of all animals to determine the incidence and types of tumors. The incidence of tumors among the experimental group was not significantly different from that occurring spontaneously among the control animals", and: "In the pathological examination of tissues from dogs on long-term feeding studies, there have been no reports of animals showing tumor incidence significantly greater than that found in the control animals".

Kettering Laboratory (in press) repeated with endrin the Fitzhugh *et al.* (1964) and Tunstall (Walker *et al.*) studies with dieldrin, and found neither hepatomata nor an increase of any other tumor after 2 years of exposure at dietary levels of 0.3 and 3.0 ppm.

6.5   METABOLISM AND BIOCHEMISTRY

According to the earlier studies:
a. Endrin, when fed to animals, shows very little tendency to be stored unchanged in the tissues, particularly fatty tissues (Kiigemagi *et al.* 1958, Street *et al.* 1957, Terriere *et al.* 1958, Treon *et al.* 1955).

b. Endrin is excreted in milk and eggs if fed at high levels (Ely *et al.* 1957, Street *et al.* 1957, Terriere *et al.* 1958).

c. Data on the storage ratio (ratio of level in fatty tissues/dietary level) are not always in good agreement, but most well-documented studies indicate values, depending on the dietary level (Terriere *et al.* 1958, Kiigemagi *et al.* 1958, Treon *et al.* 1955).

d. The biological half-life in cattle and sheep, calculated from the data contained in these publications, varies from 1.8-8.2 weeks (Robinson 1962).

Recent studies on rats by Ludwig *et al.* (1966) and Korte (1967, 1968) receiving single and repeated oral, or single intravenous doses of endrin-^{14}C and other studies by the same authors lead to the following conclusions:

1. After oral administration in peanut oil, endrin is rapidly absorbed, metabolized and excreted via the faeces (less than 1% of the administered activity is excreted in the urine).

2. When endrin is administered at lower dosages to female rats a larger percentage is metabolized and excreted than at higher dosages.

Thus the half-life of endrin, administered at a dose of 16 μg/kg is about 2 days, when administered at a dose of 128 μg/kg about 6 days for rats of both sexes.

3. In a feeding experiment with 3 rats of each sex, receiving daily doses of 8 μg/kg endrin for 12 days, male animals excreted 60% of the daily administered activity the first day, female animals 39%. The steady state of storage was reached 5 - 6 days after the first administration. 24 hours after the last administration male rats had stored only 14% of the total amount administered, the females 28%; three days later these values had dropped to 5.3% and 15% respectively.

4. The excreted faecal activity during the period of administration consisted of 70 - 75% of hydrophilic metabolites, the remainder was unchanged compound, 24 hours after cessation of dosing only hydrophilic metabolites were excreted.

5. The excretion-curve, measured in 2 rats of each sex receiving a single i.v. injection of 200 μg/kg endrin- ^{14}C shows an exponential slope. The half-life measured from this experiment is 2-3 days for males and 4 days for female rats. Metabolites could be detected in the excreta, but no unchanged endrin was found.

6. The metabolites consisted of more than 95% of a hydrophilic product, which on gas liquid chromatography examination was identical with the keto-rearrangement product of endrin such as occurs on heating or UV irradiation, and less than 5% of an even more hydrophilic compound.

7. Conversion of endrin to its metabolites occurs in the liver. 100 g pig liver in Krebs-Ringer solution with 45 mg NADPH$_2$ converted, under suitable conditions, 40% of 38 μg endrin within 72 hours to hydrophilic metabolites identical to those found in living organisms.

8. Insect larvae metabolize endrin to more hydrophilic products, one of which is on GLC examination, identical to the "endrin-ketone".

9. Cabbage plants metabolize endrin; one of the two metabolic products found is identical with this "endrin-ketone". The other has not yet been identified.

10. Fungi (*Aspergillus flavus*) metabolize endrin to products similar to those found in cabbage plants.

Cole *et al.* (1968) dosed male rats, with and without bile fistulas with intravenous doses of 0.25 mg/kg endrin-^{14}C or dieldrin-^{14}C. The urine and faeces were collected daily. The bile was collected 1, 3, 6, 12 and 24 hours after injection and then at daily intervals. After 5 - 7 days the animals were sacrificed. The radioactivity in all excreta, bile and tissues was measured by liquid scintillation counting. Over 90% of the excreted activity was found in the faeces of the intact animals and in the bile of the animals with a bile fistula. 50% of the endrin administered was excreted within 24 hours and 50% of the dieldrin within 3 days. Of the endrin administered 50% was excreted in the bile within 1 hour, whereas 32% of the dieldrin administered had been excreted in the bile after 6 hours. The authors state that the more rapid elimination of endrin as compared with dieldrin can be explained by its more rapid biliary excretion. Intestinal absorption and fatty tissue sequestration were excluded as major factors, and the liver was identified as the source of the difference.

These experiments confirm Korte's data (1967) and show that endrin is metabolized more rapidly, excreted more rapidly, and stored to a much smaller extent than dieldrin.

The endrin precursor isodrin is rapidly epoxidized to endrin by livermicrosomes in the rat (Wong and Terriere 1965).

Nelson et al. (1956) reported an elevation of the serum alkaline phosphatase level in rats fed various levels of endrin: 1, 5, 25, 50 and 100 ppm. A critical review of their paper shows that significant elevation occurred at the 25 and 50 ppm level. The assumption that the demonstrated serum alkaline phosphatase elevation is related to liver damage is not confirmed by other biochemical and histological data.

In our opinion hepatic enzyme induction at these high levels of feeding is a more likely explanation for this elevation of serum alkaline phosphatase. This would be analogous to a similar isolated finding in dogs at the highest dose level in a long-term feeding study with dieldrin (Walker et al. 1968).

6.6   PHARMACODYNAMICS

In the rat experiments by Korte (1967), the distribution of radioactivity in the organs and tissues was measured 25 hours after intravenous injection and 24 hours and 4 days after cessation of a 12 day period of oral dosing. Storage occurs mainly in abdominal and subcutaneous fat and to a lesser extent in liver, brain, muscles and other organs.

In an experiment by Richardson et al. (1967) 3 beagles received a daily dose of 0.1 mg/kg (equivalent to approximately 4 ppm) endrin for 4 months. Blood samples were collected weekly, 7 days after the final administration the dogs were sacrificed and samples of blood, body fat and other selected tissues analysed for endrin content. No endrin was found in the blood before initiation of the experiment. The concentration of endrin in the blood remained stationary from the second day of the experiment, varying only slightly from 0.0020-0.0084 μg/ml. At these maximum levels of endrin in the blood there were no symptoms of toxicity. DDT + DDE levels in the blood remained at essentially the same levels of about 0.002 μg/ml throughout the experiment. At the conclusion of the experiment the highest concentration was found in the adipose tissue and this concentration in the fat varied between 0.250 and 0.760 ppm and it was related to the level in the blood.

In a simultaneous experiment on dogs, dosed with an equal amount of dieldrin, storage-equilibrium in the blood was reached after 114-121 days at levels varying from 0.097-0.190 ppm and at autopsy the levels in the fat varied from 12-44 ppm.

6.7 DISCUSSION

In long-term feeding studies there is some disagreement between the results of Nelson *et al.* (1956) and the later studies. On a closer examination the measurable changes in the former study, such as the stated increase in serum alkaline phosphatase and the decrease in body weight, were only significant at the 25 and 50 ppm dietary levels. Based on more recent prolonged feeding experiments a 1 ppm dietary level may be regarded as the no-effect level in the rat and the dog.

Although, as far as we know, no specific studies on hepatic enzyme induction have been done, this no-effect level is also applicable to enzyme levels studied and DDT + DDE levels determined (Richardson *et al.* 1967).

Although the acute toxicity of endrin is about 4 to 5 times greater than that of dieldrin, the dietary no-effect level is 1 ppm for endrin. This may be explained by the fact that endrin is more rapidly metabolized and excreted by vertebrates, and consequently stored to a very much lesser extent in the body tissues. Blood and adipose tissue levels of endrin at the state of equilibrium are thus much lower then for the same daily dose of dieldrin.

Although at a much lower level than with dieldrin, a storage plateau is also reached when endrin is fed at a fixed daily dose (Zavon 1961, Richardson *et al.* 1967).

6.8 SUMMARY

Endrin is the stereoisomer of dieldrin, but it differs from dieldrin in the following respects:
– it has a higher acute toxicity,
– it is more rapidly metabolized in plant and animal tissues,
– endrin is not a persistent insecticide in vertebrates.

Symptoms of intoxication due to stimulation of the central nervous system are similar to those caused by dieldrin. The LD_{50} varies with the vehicle and the experimental species used, but on an average endrin is 4 to 5 times as acutely toxic as dieldrin.

The dietary no-effect level for endrin in chronic feeding studies in the rat and the dog is 1 ppm. Endrin is not a carcinogen.

The pharmacodynamics of endrin are similar to those of dieldrin, although the storage levels reached at the same dietary dose are much lower than with dieldrin.

Endrin is rapidly metabolized to hydrophilic metabolites which are excreted, for more than 90% appearing in the bile and the remainder in the urine.

84

Chapter 7

General toxicology of endrin — studies on man

7.1   INTOXICATION

7.1.1 Cases of intoxication in the medical literature

In one incident 59 people became ill from the ingestion of bread accidentally containing up to 150 ppm of endrin, but there were no fatalities (Davies and Lewis 1956). Calculations based on the amount of bread consumed suggested that an intake of 0.2-0.25 mg/kg could produce a convulsion (Hayes 1963), whereas the maximum amount consumed was estimated to have been 1 mg/kg (Zavon 1961).

Data on the pathology of 60 fatal cases of endrin poisoning, 41 of which involved suicide, have been published by Reddy *et al.* (1966). No specific histopathological organ changes were observed. These data do not include any reference to the size of the doses ingested. However, the toxic doses of endrin were estimated to be 5-50 mg/kg by the oral route, the lethal dose of endrin was estimated to be about 6 gram (Reddy *et al.* 1966).

Hayes (1963, 1966) states that endrin has been found at concentrations of up to 400 mg/kg in the fat and up to 10 mg/kg in other tissues of people fatally poisoned by it.

Van Raalte (1965) extracted from the world literature all cases of fatal endrin poisoning known at the time: a total of 97 cases were reviewed of which 69 cases were suicide, 24 accidental ingestion, 4 occupational cases from endrin spraying and none as a result of endrin manufacture and formulation.

Coble *et al.* (1967) reported 3 cases of non-fatal convulsive endrin poisoning in the United Arab Republic resulting from the consumption of bread made from endrin-con-taminated flour. In the first case the serum endrin level was 0.053 µg/ml, 30 minutes after the convulsion endrin could not be detected (<0.004 µg/ml) in samples of the cerebrospinal fluid taken at the same time. 20 hours after the onset of convulsions the serum endrin level had fallen to 0.038 µg/ml, 10 hours later still, it was 0.021 µg/ml. In cases 2 and 3 no endrin was detected in the blood 8½ and 19 hours respectively after convulsions.

Weeks (1967) described 4 major outbreaks of acute endrin poisoning which occurred in Quatar and Saudi Arabia during June and July 1967. 874 persons were hospitalized and 26 died. All victims had eaten bread made from flour contaminated with endrin during transport, which contained from 2000-4000 mg/kg endrin. Blood from patients contained 0.007-0.032 µg/ml endrin.

7.1.2 Symptomatology, treatment and prognosis of intoxication

Frequently the first indication of acute endrin poisoning is a sudden epileptiform convulsion, occurring from 30 minutes to up to 10 hours after exposure (Weeks 1967). It lasts for several minutes and is usually followed by a semi-conscious state for ¼–1 hour (Coble et al. 1967). Death, or permanent brain damage, may ultimately occur as a result of anoxemia due to prolonged convulsions (Jacobziner et al. 1959, Coble et al. 1967). In less severe cases of endrin poisoning the primary complaints are headache, dizziness, abdominal discomfort, nausea, vomiting, insomnia, agressiveness and, rarely, slight mental confusion (Coble et al. 1967, Weeks 1967). The prognosis is good if cerebral damage by prolonged anoxemia is avoided. Recovery to full normal health in such cases is rapid and usually complete within a few days (Davies et al. 1956). No specific findings from acute endrin poisonings have been reported at autopsy (Reddy et al. 1966, Coble et al. 1967, Weeks 1967). The rapidity of the onset of signs and symptoms, predominantly of central nervous system stimulation, and the rapid return to normal among those who survive, is typical for an intoxication with an organochlorine insecticide (Weeks 1967). The recovery from an endrin intoxication is quicker than that from the other insecticides under discussion.

Therapy for endrin intoxication is the same as that for aldrin and dieldrin intoxication and is described in 5.1.3.

7.2 PHARMACODYNAMICS AND BIOCHEMISTRY

No human volunteer study has been conducted with endrin. However, all data available indicate that endrin is very similar to aldrin and dieldrin (as described in 5.2) with the major exception that, as referred to in the animal studies (6.5 and 6.6), it is much more rapidly metabolized and excreted. This also applies to man. Thus it is stored to a much lesser extent than dieldrin in the tissues, in fact, in human tissues endrin is detectable only at a very low level, and then only shortly after acute over-exposure.

7.3 EXPOSURE OF THE GENERAL POPULATION

7.3.1 Measured daily intakes

When compared with those on dieldrin, data on endrin concentrations in water and food, are very sparse.

Endrin was not detected in the water of the lower Mississippi (Novak et al. 1965) which drains an area where endrin is extensively used for the control of agricultural pests, nor has it been detected in streams and waterways in Great Britain (Report by the Advisory Committee, the "Wilson Committee" 1969).

In Great Britain endrin could not be detected in rain water at 7 different locations at a detection level of 1 part per 10^{12} during 1966–1967 (Tarrant et al. 1968).

Studies on complete prepared meals in the U.S.A. and Canada showed that endrin occurs only exceptionally, and then only in trace amounts (Williams 1964, Cummings 1965, Campbell et al. 1965, Duggan 1966).

In complete prepared meals in the U.K. no endrin was found during surveys conducted in 1965 and 1967 (Robinson et al. 1965, Robinson et al. 1966, McGill 1969).

7.3.2 Endrin levels in adipose tissue and blood

In most of the surveys of insecticide levels in adipose tissue and blood (as tabulated in chapter 5.4.2 and 5.4.3) samples were examined for endrin as well. Even in those areas where endrin is most extensively used (*e.g.* India and Lower Mississippi area) endrin could not be found in human subcutaneous fat or in blood from the general population at the limit of detection of 0.03 mg/kg and lower (Kunze *et al.* 1953, Hoffman *et al.* 1964, Dale *et al.* 1965, Zavon *et al.* 1965, Novak *et al.* 1965, Robinson *et al.* 1965, Wiswesser 1965, Hayes *et al.* 1965, Brown 1967, Hayes *et al.* 1967, Wasserman *et al.* 1968, Hayes *et al.* 1968, Robinson 1969).

7.4 OCCUPATIONALLY EXPOSED WORKERS

In Treon's review of the toxicology of endrin (1956) he states: "These studies (on workers handling endrin) reveal that harmful physiological effects to workers are found only in those instances where excessive absorption has occurred, either in the form of an acute dose or subacute doses from unusually careless handling. No established cases of chronic illness from exposure to endrin are on record".

In a manufacturing plant in the U.S.A., medical supervision of workers exposed for a period of 1-19 years (average 12 years) failed to reveal any adverse effects (Hayes *et al.* 1967). In these occupationally exposed workers no endrin could be found in the subcutaneous fat. In cases of non-fatal endrin intoxication blood levels fall below the level of detection a few days after cessation of contact. Once again, these observations suggest a rapid excretion of endrin in man.

Van Dijk (1968) examined serum alkaline phosphatase levels of 15 endrin operators in November 1964, July 1965 and February 1966 respectively. Some of these operators were working in the endrin plant for periods up to 8 years. No significant change in the alkaline phosphatase was found, and all levels remained normal.

7.5 DISCUSSION

From all the cases of endrin intoxication described in the medical literature, the general impression is obtained that when convulsions occur in cases of acute intoxication they are mostly without prodromal signs. Furthermore in non-fatal cases recovery from endrin intoxication is a matter of one or a few days, as compared with dieldrin where sometimes a few weeks are needed for complete recovery.

This is in agreement with the fact that after endrin intoxication blood levels fall to below the detection level within a few days.

It may be concluded from this, that endrin is a highly toxic, but non-persistent organochlorine insecticide in man, the only hazard being from acute over-exposure. When, however, sufficient preventive measures are taken during manufacture, transport and in the course of spraying, the hazard in using endrin is small.

7.6 SUMMARY

Many more cases of fatal intoxication have occurred with endrin than with aldrin and dieldrin. Most fatal cases have resulted from suicide or eating bread made from endrin-contaminated flour. This also confirms that the acute toxicity of endrin is rather high, when compared with that of aldrin and dieldrin. Data on the estimated toxic dose of endrin in man are imprecise.

Symptomatology and therapy in endrin intoxication are the same as with dieldrin intoxication. If cerebral injury resulting from prolonged anoxemia due to convulsions does not occur or is prevented, full recovery to normal health may be expected within days.

Residues of endrin are not normally found in water or food.

Endrin is not stored in the body fat of the general population and occupationally exposed workers. It is a non-persistent insecticide in man.

Chapter 8

General toxicology of Telodrin

8.1   ACUTE TOXICITY AND SYMPTOMATOLOGY

A summary of acute toxicity data from animal tests is given in Table 8.

There appears to be no appreciable difference between the sexes in their susceptibility to single exposures. The toxicity of a given dose of Telodrin varies with the vehicle, the route of administration and the test animal employed. Wettable powder formulations resemble the technical product in toxicity, but in xylene emulsions it is more toxic both by oral and dermal routes. The presence of solvents in Telodrin formulations appears to promote absorption through the skin, even though crystalline unformulated Telodrin itself, is readily absorbed by this route.

Table 8

ACUTE TOXICITY OF TELODRIN IN EXPERIMENTAL ANIMALS

Animal species	Way of entry	Formulation	LD_{50} in mg/kg body weight	Reference
HLS-rat	oral	xylene emuls.	3.24	1
id.	id.	water emuls.	5.36-5.60	1
id.	id.	corn oil	10.7	2
CF-rat	id.	water emuls.	7.14-7.38	1
CF-mouse	id.	water emuls.	6.31	1
Chicken	id.	xylene emuls.	2.7	1
id.	id.	water emuls.	3.7	1
id.	id.	corn oil	2.0	2
Guinea pig	id.	arachis oil	2.0-3.0	2
Rabbit	id.	arachis oil	4.0	2
Dog	id.	arachis oil	1.0-2.0	2
HLS-rat	dermal	xylene emuls.	20-22.5	1
id.	id.	techn. Telodrin	60	1
id.	id.	arachis oil	4-10	2
Guinea pig	id.	arachis oil	2.0-10.0	2
Rabbit	id.	arachis oil	2.0-7.5	2
CF-rat	intraperiton.	water emuls.	3.56	1
CF-mouse	id.	water emuls.	8.17	1

HLS = Hooded Lister Strain; CF = Carworth Farm.
References: 1 = Brown *et al.* 1962; 2 = Worden 1968.

The symptoms noted during Telodrin intoxication are similar to those seen in intoxication by aldrin, dieldrin or endrin. Although there are minor differences between the effects of these insecticides in different species, it is generally impossible to distinguish between the toxicants on the grounds of symptomatology alone (Brown *et al.* 1962, Worden 1968, Zavon 1961).

The symptoms of intoxication vary slightly with the different species of experimental animal. The onset of symptoms depends on the dose, the route of administration and the formulation in which the Telodrin is administered.

Increased irritability, tremors and tonic-clonic convulsions in that order are the major symptoms: all of them indicating that the principal site of action of Telodrin is the central nervous system (Brown *et al.* 1962, Zavon 1961, Worden 1968).

Symptoms begin to appear 1-2 hours after the administration of a sufficiently large acute oral dose. In some animals, *e.g.* rabbits, dyspnoe may be the first symptom.

The majority of deaths in rats occurred within 20 hours of the administration of the oral test dose. Those which recovered − even after violent convulsions had occurred − did so rapidly and were normal in appearance and behaviour within a few hours.

Brown *et al.* (1962) fed groups of eight adult male and eight adult female Carworth Farm Rats at dietary levels of 0.25, 1.0 and 2.5 mg/kg Telodrin daily: 5 days per week for 2 weeks. At the 2.5 mg/kg level all animals were dead after the 5th dose; at the 1.0 mg/kg level only 6 male rats survived all 10 doses; at the 0.25 mg/kg level only one female rat died after the 5th dose. No rats died during the 10 days following the experiment. Animals dosed at the highest level lost weight prior to death; this effect did not occur at the lower dose levels.

The same authors injected groups of 5 male and 5 female adult Carworth Farm Rats with intraperitoneal doses of 0.25, 1.0 and 2.0 mg/kg/day of Telodrin (as a 1.0% w/v solution in dimethylsulphoxide) 5 days per week for 2 weeks. At the 2 mg/kg level only one female rat survived, at the 1 mg/kg level 3 male and 3 female rats survived and at the 0.25 mg/kg level all rats survived.

Consideration of the results from both tests suggests that the apparent additive effect of a few repeated oral doses and the sex difference on repeated oral exposure, are, in fact, related to the rate of absorption in the body, rather than to any increased susceptibility of the animal or inherent sex difference in susceptibility.

8.2   SUBACUTE AND CHRONIC TOXICITY

Worden (1968) during tests carried out from 1957-62 at the Huntingdon Research Centre, fed six groups of 25 male and 25 female Hooded Lister Rats at dietary levels of 5, 17.5 and 30 ppm Telodrin for two years. Three of the groups were controls. The 30 ppm level in the diet is approximately 1/5th of the acute LD_{50} dose for this species.

From the first day some rats at the 30 ppm level showed signs of increased irritability of the central nervous system with occasional convulsions; these effects however abated after the first few weeks of study; from the 3rd month onward only an occasional convulsion occurred.

At the 17.5 ppm level milder symptoms and an occasional convulsion occurred during the first few weeks only, with no observable symptoms thereafter.

No toxic symptoms, attributable to Telodrin, were seen at any time in the rats on the 5 ppm level, the general health of these rats being at least equal to those of the untreated control rats.

Apart from the 5 female rats on the 30 ppm level which died in the early stages of the test, there was no significant effect from Telodrin on mortality at any level. There was no adverse effect from Telodrin on the body weight at any of the levels, in fact all three male test groups had a higher weight gain than the controls. Telodrin administration did not produce an adverse effect in the differential white cell count nor in the liver function tests, at any of the three levels. The liver and kidney weight of the 30 ppm female rats were increased as compared with the mean control value, but not as compared with the particular control group at that level. Histologically the only change which might possibly have been associated with Telodrin was that of thyroid hyperplasia which was recorded in 6 females and 4 males at the 30 ppm level.

A dietary level of 5 ppm was clearly a no-effect level under the conditions of this experiment.

Brown et al. (1962) fed groups of 20 male or female Carworth Farm Rats at dietary levels of 10, 5, 1 and 0.1 ppm Telodrin for 100 days. Only at the 10 ppm level were fatalities significantly higher than in the control group. Again 5 ppm was the no-effect level in this trial.

Ware et al. (1967) when using Balb/c mice, found that a 10 ppm dietary level of Telodrin was lethal within 24 days; a 5 ppm dietary level produced a 100% mortality by the 64th day; 2.5 ppm level produced 80% mortality after 120 days. At a 1 ppm dietary level all mice survived normally. 1 ppm was the no-effect level in this test.

Worden (1968) between 1957 and 1962 treated three groups of 3 male and 3 female Scottish Terriers with 0.0, 0.5 and 2 ppm Telodrin corresponding with 0.025 and 0.1 mg/kg/day respectively for two years by gastric tube. There were no deaths associated with the compound, nor was there any evidence of an effect from Telodrin on food consumption, body weights, haematology, blood chemistry, SRE, differential white blood cell count, or routine urinalysis. However, liver, kidney and lung weights were significantly higher in the males receiving 0.1 mg/kg/day. At this same level there was a slight elevation of the serum alkaline phosphatase. There was no evidence of any histological change resulting from exposure to Telodrin in any of the tissues examined. Telodrin concentration was highest in the body fat and least in the blood. There was a significant correlation between the levels in the blood and that in other tissues. The 0.025 mg/kg/day dose level was the no-effect level under the conditions of the experiment.

Brown et al. (1962) dosed beagles, one male and one female to a group, 7 days a week with doses of 0.08, 0.125 and 0.2 mg/kg/day (being equivalent with dietary levels of 1.6, 2.5 and 4.0 ppm) for 3 months. All dogs at the 0.2 and 0.125 mg/kg/day dose developed convulsions when the Telodrin concentrations in their blood ranged from 0.042-0.072 μg/ml. Those on 0.08 mg/kg/day and the control dogs showed no adverse effects.

The same authors (1962) performed autopsies on all rats and mice which died during the above mentioned studies, both from obvious intoxication as well as other causes. Thus 916 rats and 56 mice were examined. The exposure of rats and mice to Telodrin in lethal and sub-lethal doses was not found to cause structural changes in any organs or tissues examined. Pathognomonic signs indicative of Telodrin intoxication were absent. The exposure to Telodrin did not affect the course of the indigenous disease of these species. From the evidence available, the authors conclude that exposure to Telodrin does not result in structural injury in living tissue and that death is due to, or associated with, functional injury in the central nervous system.

We know of no enzyme induction-tests that have been carried out with Telodrin.

Dermal irritation tests

Worden (1968) described the results of dermal tests in guinea pigs, rabbits and rats. All tests were negative, there being no evidence of irritation or visible changes, following application of Telodrin to the skin for up to 30 successive days.

8.3 CARCINOGENESIS

According to Stevenson (1964): "There has not been any evidence in the chronic toxicity studies of an increased incidence of tumours in the animals receiving Telodrin, compared with the control groups".

8.4 METABOLISM, BIOCHEMISTRY AND PHARMACODYNAMICS

Korte (1963, 1967) administered 7 μg Telodrin -^{14}C per animal intravenously in male and female rats. In the faeces and urine of both sexes a hydrophilic metabolite, a lactone, was excreted as the main metabolite, with only traces of unchanged Telodrin. The excreted lactone had a mammalian toxicity 30 times less than that of Telodrin. Within 48 hours after injection 12% of the activity administered was excreted in the faeces and 1-5% in the urine. In rabbits, however, as in the case of dieldrin, the major portion of radioactivity was excreted with the urine.

The small amounts of Telodrin found in liver, heart and blood 48 hours after injection as compared with the concentration in other tissues, indicate that Telodrin is initially stored in the fat and then slowly excreted.

Moss and Hathway (1964) found that the solubility of Telodrin in rabbit serum is 4000 x as great as the solubility in water. Absorbed Telodrin is primarily located in the erythrocyte contents and the plasma of the rabbit and rat blood and not in the leucocytes, the platelets or the erythrocyte stroma. The distribution of Telodrin between plasma and red cells is roughly 2 : 1, similar to that of dieldrin. In the red cells Telodrin is largely associated with hemoglobin and an unknown other constituent, in the plasma with albumin, α_1- and α_2-globulin and another unidentified component, as Moss and Hathway found in experiments with Telodrin -^{14}C in rabbits. The erythrocyte surface is freely permeable to Telodrin.

Hathway et al. (1967) were able to demonstrate the likelihood of a two-way transplacental passage of Telodrin in rabbits.

Hathway et al. (1964) dosed CF-rats with 7.5 mg/kg intraperitoneally and sacrificed them at different times before and after convulsions. As the pharmacological effects of Telodrin relate to the central nervous system, biochemical examinations were confined to the brain. From the results it was concluded that Telodrin causes liberation of ammonia in the brain and that this occurs prior to the onset and throughout the course of convulsions. This is preceded by an overwhelming of the ammonia-binding mechanism of glutamic acid, glutamine and α-oxoglutaric acid, furthermore, there is a very rapid pre-convulsive synthesis of cerebral glutamine. Some animals treated preventively with 50 mg/kg glutamine were resistant to Telodrin convulsions.

In the chronic exposure tests by Brown *et al.* (1962) determination of the serum alkaline phosphatase, aliesterase and cholinesterase activities showed no abnormalities during the test period.

8.5 CENTRAL NERVOUS SYSTEM AND PERIPHERAL MOTOR EFFECTS

Chambers (1962) recorded the electrical activity of the brain of the anesthesized cat before and after slow intravenous injection of Telodrin in 5 and 10 mg/kg doses. Spikes were seen in the pre-convulsive state and they were present in excess as synchronous high voltage spikes during the convulsive episodes.

The onset of convulsive seizures in experimental animals could rapidly be controlled by intravenous injection of pentobarbital sodium at a dose of about 30 mg/kg body weight. Atropine, however, enhances the toxic effect of Telodrin as well as that of dieldrin (Chambers 1962).

Ibrahim (1964) showed that Telodrin injected intraperitoneally in sub-lethal or lethal doses into male Wistar rats produced a higher tension of contraction in the gastrocnemius muscle at lower frequencies of stimulation than in controls; the maximum tetanic tension was also attained at a lower frequency. She considered the increase in the duration of the active state of the muscle to be the most likely explanation for this effect.

8.6 HUMAN EXPOSURE DATA

As far as we are aware, no cases have been reported in the literature in which intoxication or skin irritation has occurred among men engaged in manufacturing, formulating or otherwise working with Telodrin.

Based on results obtained in animal tests, Brown *et al.* (1962) expect in case of human intoxication, predominant signs of central nervous system stimulation, with headache, irritability, nausea and dizziness; possibly followed by convulsions and perhaps preceded by anorexia and weight loss.

No data were found in the literature on the average exposure of the general population to Telodrin or on occupationally exposed workers.

8.7 SUMMARY

The acute oral toxicity of Telodrin for a variety of mammals was found to be between 1.6 and 10 mg/kg (LD_{50}). Symptoms of intoxication are stimulation of the central nervous system, closely resembling those of intoxication by aldrin, dieldrin and endrin.

In dermal exposure tests Telodrin did not produce irritation or visible changes when applied to the skin of rats, guinea pigs and rabbits.

The repeated oral exposure of rats indicated that when 1/3rd-1/7th of a single LD_{50} dose was given daily, these doses were more or less additive. These results suggest that a high degree of accumulation was occurring, which seemed to be related to the rate of absorption in the gut.

In chronic toxicity studies a no-effect level was found at a dietary exposure of 5 ppm in rats, 1 ppm in mice and 1.6 ppm in dogs. Telodrin did not cause any histological changes in tissues or organs at lethal or sub-lethal levels, nor did it affect the course of indigenous disease. Only at the 30 ppm dietary level was a thyroid hyperplasia found in rats.

The main effect of an overdose of Telodrin is a disturbance in the central nervous system, with convulsions. This CNS effect is reversible.

Barbiturates are effective for the control of convulsions.

After its metabolization to a less toxic and more hydrophilic lactone, excretion takes place in urine and faeces.

Convulsions due to Telodrin are probably caused by liberation of ammonia in the brain after overwhelming of the ammonia binding mechanism.

Telodrin is transported in the blood, bound mainly to haemoglobin, albumin and α_1- and α_2-globulin.

Human toxicity data have not been reported.

C. LOCAL SITUATION

Chapter 9

Brief description of the insecticide plant

9.1 PRODUCTION HISTORY OF CHLORINATED HYDROCARBON INSECTICIDES AT PERNIS

The production of aldrin was started in November 1954 and was followed by that of dieldrin in June 1955. At the same time a production plant for intermediates and a formulation unit for the production of commercial products from the technical insecticides was started up.

A separate unit for the production of endrin was added in February 1957, since which time these three insecticides have been manufactured and formulated.

A pilot plant for the production of Telodrin was added to the aldrin-dieldrin plant and was in operation from July 1958 until September 1961, when a separate Telodrin plant started production. Telodrin production was discontinued in September 1965. However, formulation and handling of Telodrin continued for some time.

Initially, products in all plants were manufactured batch-wise. Later aldrin and Telodrin were made by a continuous process. Dieldrin and endrin still are produced batch-wise.

Since 1954 together with new developments, many improvements in the technical equipment and industrial hygiene have been made. This development continues, with closed circulation and automation being introduced wherever possible.

From 1964 onwards the routine determination of insecticide levels in the blood became the principal method of indicating danger spots and unsafe handling. See also chapter 13.

9.2 GENERAL LOCATION AND PLANT UNITS

9.2.1 *General situation*

The insecticide plant of Shell Nederland Chemie is located amongst other chemical and oil refinery plants, and closely interlocked with them in the vast industrial site which

Shell Nederland Chemie and Shell Nederland Raffinaderij occupy at the first and second oil-harbour at Pernis-Rotterdam.

The industrial medical department is located within the gates of this complex not far from the insecticide plant.

The insecticide plant is subdivided into 4 separate units, each with a separate team of personnel:

a. intermediates plant,
b. aldrin-dieldrin production plant,
c. endrin production plant,
d. formulation unit

For some time there was a 5th subunit:

e. Telodrin production plant.

A separate department is the

f. filling and despatching department.

In this context special mention should be made of

g. laundry for work-clothes of insecticide workers.

Intermediates, aldrin and dieldrin plant, and the former Telodrin plant, are open units with a control-room attached. The endrin plant, formulation unit and the storage areas are closed buildings.

It is outside the scope of this study to discuss in detail the manufacturing process and the toxicological details of intermediates and other chemicals used in this process.

9.2.2 *Formulation unit*

In our formulation unit various types of formulations are produced. Originally only organochlorine insecticides were formulated, but for some years, and in increasing amounts, organophosphates, as well as mixtures of the two, have also been formulated.

Solid formulations include:

a. wettable powders,
b. dust concentrates,
c. field strength dusts,
d. granules.

Liquid formulations include:

e. emulsifiable concentrates,
f. solutions.

Wettable powders and dusts are prepared by dry blending and milling toxicant with inert mineral carrier fillers and some other ingredients, for instance emulsifiers or pigments. Granules are prepared by impregnation or pelletizing.

Liquid formulations are produced batch-wise at atmospheric pressure and at temperatures ranging from ambient to $40°C$, according to the quantity of the active material to be dissolved and the type of solvent to be used.

Solid formulations a. and b. require varying degrees of milling to reduce the particle size of the product to specification. The highest milling intensity is needed for category a. products, which successively pass a hammer-mill and a fluid-energy mill. For category b. products a hammer-mill is sufficient. Category c. is made by dry blending of a dust concentrate b. and a milled filler. Category d. products do not require any milling. Since

December 1967 there has been a special granulation unit in the formulation plant.

Mixing vessels with circulating pumps and filters are used in the production of the liquid formulations.

As many sources for air and water pollution exist in this plant, extensive preventive measures have been taken. These will be described in the separate sub-chapter 9.4.

9.2.3 Filling and despatching department

This is a separate department, which is responsible for filling, storing and despatching products of these and of many other chemical plants.

Here packing and labelling according to national and/or international standards is carried out.

Solid formulations are drummed off in the formulation unit, liquid formulations in a separate part of the liquid-filling installations.

Technical and formulated insecticides are stored in two closed storage buildings of this department and products are despatched as required.

9.2.4 Insecticide laundry

Since October 1957 a special laundry for work-clothes of insecticide workers has been in use next to, but incorporated in, the insecticide plant. This laundry is equipped with:

3 Reineveld washing machines,

2 centrifuges,

2 drying apparatus, and since September 1964 with:

1 dry cleaning machine.

This laundry is operated in 4 shifts with extra personnel during day-time.

A description of the methods and procedures in this laundry will be given in the sub-chapter about personal hygiene measures of the next chapter.

9.3   SAFETY REGULATIONS AND PROVISIONS IN THE INSECTICIDE PLANT

A personal loose-leaf booklet containing all general safety regulations for all plants at our site is issued to everyone working with Shell Nederland Raffinaderij and Shell Nederland Chemie. Instruction is given in these rules in the instruction courses at our Industrial Technical School including demonstrations with, for instance, static electricity, explosive mixtures, etc., as well as in the use of the safety devices available. Smoking is prohibited on all plant sites. To be found smoking in these places means prompt dismissal.

Every worker has to know these regulations and adhere to them: safety-mindedness is one of the main points in yearly reviews of staff reports.

In addition to these general safety rules every worker in the insecticide plant has an extra set of safety regulations which apply specifically to that plant. These special regulations consist of three parts:

1. General safety rules for the insecticide plant, which will not be discussed here as they are not relevant in this context.

2. Regulations concerning plant hygiene.

3. Regulations concerning personal hygiene, which will be discussed in the next chapter. The most important rules concerning plant hygiene are:

— Access to insecticide plants and storage areas is restricted to persons working there. Others have to get special permission to do so.

— Keep the plant in an immaculate state of cleanliness.

— Keep door handles, hand rails and all equipment clean.

— Contaminated equipment or safety materials must be delivered to the insecticide laundry to be decontaminated.

— Avoid making dust, especially when working with toxic material.

— Spills of liquid toxicants must be covered with inert absorbant material, swept up and disposed of in the specially marked dust-bins to be burnt in the solid waste incinerator.

— All waste material must be disposed of in above mentioned marked dust-bins to be burnt in the solid waste incinerator. Used gloves must be cut before depositing them in these bins.

At every place in the plants where gross accidental contamination with insecticides might occur safety showers operated by hand or by foot are available.

9.4   WASTE DISPOSAL AND PREVENTION OF AIR AND WATER POLLUTION

The following aspects will be discussed briefly:
1. Liquid waste disposal.
2. Solid waste disposal.
3. Prevention of air pollution.
4. Prevention of water pollution.

9.4.1 Liquid waste disposal

Liquid waste from equipment used in manufacture or formulation of insecticides consists of flush-solutions, scrub-liquids or residues consisting of combustible solvents.

This waste is disposed of by burning at our site mixed with fuel-gas in a liquid incinerator which is provided with a water scrubber for the removal of corrosive or toxic flue gases (*e.g.* HCl, free chlorine or phosgene) in order to prevent air pollution.

9.4.2 Solid waste disposal

All solid insecticide-contaminated waste is burned in the solid waste incinerator at our site, which is also provided with a water scrubber for the removal of corrosive and toxic flue gases (HCl, free chlorine and phosgene) in order to prevent air pollution.

All solid waste materials from our insecticide plant, as well as liquid wastes absorbed in inert absorptive substances and spent precoat of the water purification filter, together with old shoes, gloves and heavily contamined clothes, etc., of our insecticide workers are burned in this waste incinerator in a flame of at least 700°C fueled by fuel-gas.

Drums, pails and other non-flammable material contaminated with insecticides are heat-cleaned in this way with a residence time of at least 2 hours in the furnace before being disposed of as "ordinary" waste material.

Ashes are regularly checked for toxic material and when necessary recirculated through the incinerator.

9.4.3 Prevention of air pollution
1. Prevention of internal air pollution
— good housekeeping in general,

— the prevention of generating toxic dust, by installing dust-proof equipment as much as possible,

— local exhaust at all points where toxic dusts or vapours may escape: over charging hoppers, filling cabinets, pumps, mixers, etc. This local exhaust consists of exhaust-hoods and/or spot-exhausts (for instance at filling points) connected to a central air exhaust system,

— the installation in the formulation plant of a central vacuum cleaning system for immediate removal of powder deposits, for cleaning floors and equipment. Connections for hoses exist at all points where needed throughout the formulation unit.

2. Prevention of external air pollution
a. Process air is purified in a double system:

— product/air separation in dust collectors (bag filters),

— scrubbing of the exhaust of the dust collectors in water scrubbers calles rotoclones.

Possible slip of the bag filters is caught in this way. There is a photo electric control system between dust collectors and scrubbers. The scrubbing water is led to the water purification unit.

b. The exhaust air of the above mentioned central exhaust system passes a static bag-filter. This filter is regularly cleaned. Dust deposits are rounded up and disposed of in the solid waste disposal unit.

c. Several cabinets for filling toxicants, ingredients, etc., are exhausted directly via rotoclone scrubbers.

d. Process and exhaust air systems in the new granulation unit are provided with rotoclone scrubbers.

9.4.4 Prevention of water pollution
The following potential sources of water pollution exist in our plant:

— process water,

— equipment clean-out water,

— rinsing water (floor cleaning),

— scrubbing water from the rotoclone scrubbers.

All these streams are collected in a sump pit and stirred, after which they are led over a rotary vacuum filter provided with a perlite type of precoat. The outside clay-layer containing pollutants is continuously scraped off, the scrapings are collected and disposed of in the solid waste incinerator. The clear filtrate passes an oil-separator, where the oil layer is separated. For the absorption of toxicant dissolved in the filtrate a treater filled with active carbon is installed in the outgoing stream which, before passing into the sewer is continuously monitored for insecticide content.

9.5 SUMMARY

This chapter has given a general description of the layout of our insecticide plant, together with a description of the safety regulations, safety provisions and the provisions for waste disposal and those against air and water pollution.

Chapter 10

Exposed plant workers

10.1 PERSONNEL WORKING FULL-TIME IN THE INSECTICIDE PLANT

In Table 9 a tabulation of all personnel working full-time in the insecticide plant on January 1st 1968 is subdivided into:

a. operation units and service departments

b. shift- or nonshift-work

c. company- or contractor-workers.

The total number of people working at any one time in the insecticide plants has increased from 40 in 1954 to a maximum of 230 in 1962. The average number of workers per July 1st of each year is given in chapter 11, Table 11. Proportionally the distribution over the various categories has always been very similar to the one given per January 1st 1968.

10.1.1 Operation personnel

These men are responsible for the production process and are working in 4 rotating shifts. Each shift consists of a foreman with several operators and 1 or 2 drum fillers. In the formulation unit each shift consists of a foreman with 1 or 2 operators supervising formulators and drum fillers. Drum fillers and formulators are mostly contractor-workers.

10.1.2 Maintenance personnel

This consists of fitters, pipefitters, welders, electricians, instrument fitters and instrument technicians. They work day-time only. There are only a few contractor-workers in this group. Some of the sub-plants have their own maintenance crew.

10.1.3 Plant cleaners and clean-out crew

Every unit has one or more day-time contractor-workers as plant cleaners. For

Table 9

PERSONNEL WORKING FULL-TIME IN THE INSECTICIDE PLANT ON JANUARY 1st 1968

		Personnel on 4-shift-duty	Total	Normal day-time personnel	Total	
OPERATION UNITS	Intermediates unit	1 shift foreman + 8 men per shift	36	plant cleaners	2	
	Aldrin-Dieldrin unit	1 shift foreman + 7 men per shift	32	plant cleaners	3	
					1	day foreman
					1	assistant department manager
	Insecticide laundry	1 man	4	1 foreman + 2 men	3	
					1	department manager
	Endrin unit	1 shift foreman + 7 men per shift	32	plant cleaners	1	plant instructor
					2	
	Formulation unit	1 shift foreman + 7 men per shift	32	plant cleaners	3	
					1	day foreman
					1	assistant department manager
Total operation			136	+	19	= 155 of which 32 contractor-workers
SERVICE DEPARTMENTS	Maintenance department		–		30	
	Clean-out personnel		–		4	
	Filling and despatching department		–		6	
Total service departments					40	= 40 of which 16 contractor-workers
Grand total						195 of which 48 contractor-workers

special "clean-out" jobs there is a day-time and a shift group of clean-out workers specially trained for the cleaning of columns, reactors, tanks, etc.These groups consist of mixed company- and contractor-workers.

10.1.4 Filling, storage and despatching

Done by contractor-workers in two shifts rotating morning and afternoon on normal workdays, including Saturday mornings.

An indication of the turnover of personnel in the insecticide plant is given by a comparison of the personnel working for longer than 4 years and total personnel in the plant as per January 1st 1968 (Table 10). As can be seen from this table more than half of the crew have been working in insecticides for more than 4 years and this is true for all categories of workers in about the same percentage. As can be seen from Table 11, chapter 11 the total work population of the insecticide plant remained approximately constant from 1957 on.

Table 10
PERSONNEL IN INSECTICIDE PLANT PER JANUARY 1st 1968

Department	Total	>4 years in insecticides	%
Operation	155	82	53
Maintenance	30	19	63
Clean-out	4	2	50
Filling dept.	6	3	50
Grand total	195	106	54.4

10.2 SELECTION AND TRAINING OF PERSONNEL

Toxic materials cannot be handled safely with workers unable or unwilling to observe safety precautions (Zielhuis 1960). Therefore selection and training of personnel are very important.

10.2.1 Selection

The responsibility for selecting a safety-minded crew is primarily that of the personnel department. Afterwards there is always the possibility of transferring those workers who are not performing up to safety standards. In the earlier production years selection especially of contractor-personnel was not performed as critically and as strictly as later on, and this, together with less strict supervision of hygienic discipline resulted in higher insecticide exposures which caused some of the intoxications occurring at that time. The industrial medical department plays a role in this selection process, firstly with the pre-placement examination, and secondly because in the course of the routine blood examinations for pesticide levels, workers with unexpectedly high insecticide levels will be detected. On some occasions this might be an expression of rather careless work.

10.2.2 Training

All operation and maintenance personnel are trained at our own industrial technical school in special courses lasting from 3 months up to 2 years, depending on former training. Refresher courses are given when needed.

For every new plant or process the entire personnel again receives a special training: theoretical and practical.

Like every other plant, the insecticide plant has its own plant instructor who arranges for every worker to receive the training needed, while new personnel and contractor-workers are being advised and get training on the job.

All these training courses involve not only process instruction, but safety instruction as well, and this not in the last place.

All this results not only in a better appreciation of the toxicity hazard by all employees involved, but also in more experience and less turnover in personnel.

10.3 DESCRIPTION OF SHIFT WORK

All operation personnel work in a four-shift system. Our four-shift system consists of:

a. a morning shift from 06.45-14.45 on Tuesday up to and including the following Monday, followed by 3 days off duty

b. an afternoon shift from 14.45-22.45 on Thursday up to and including the next Wednesday, followed by 2 days off duty

c. a night shift from 22.45-06.45 on Friday night up to the following Friday morning, followed by a long weekend of 4 days.

During a full shift period of 4 weeks this schedule results in 7 days off work against 8 days for day-time workers. This day is compensated for with extra holidays.

All shift workers get a shift-work allowance of 30% of their salary.

Our insecticide workers receive no extra payments in money or otherwise as compared with other operation personnel. Only for the time spent in the compulsory shower-bath after the normal working hours, half an hour is paid as overtime per day.

10.4 PERSONAL SAFETY REGULATIONS IN THE INSECTICIDE PLANT

Apart from the general safety regulations applicable to all plants, which are described in a small loose-leaf booklet which is given to every new employee in our industrial complex of refinery and chemical plants, every plant has its own extra safety regulations based on the conditions in that particular plant. For the insecticide plant some of the more important special regulations in this context will be summarized below.

Because the products and intermediates are toxic materials, absorption into the human body must be avoided. Therefore the following precautions have to be taken:

a. Against absorption via mouth or nose

1. Do not eat in the plant, do not bring food into the plant.
2. Wear dust-filter or air hose mask wherever there is a possibility of toxic dust.
3. Wear air hose mask wherever toxic gas or vapour may escape.
4. Wash hands and face thoroughly before eating.

b. Against skin penetration
1. Wear only prescribed clothing and shoes.
2. Take a full shower-bath using soap and plenty of water before going home.
3. Clothing (all) has to be laundered daily.
4. In case of contamination clothing has to be changed at once (and laundered), a shower-bath has to be taken and the industrial medical department has to be contacted. In urgent cases the company ambulance has to be called for at once.
5. Shoes have to be kept in perfect condition and always worn fully laced up. When shoes are contaminated on the inside same procedure as under 4., but shoes have to be replaced and destructed in the solid waste incinerator.
6. Plastic gloves have to be worn in all places where insecticide contact might be possible.
7. Wash hands also before (and after) using the toilet.
c. All rules on clothing, washing and eating have to be strictly adhered to.

10.5   PERSONAL HYGIENE IN THE INSECTICIDE PLANT

The rules and regulations for personal hygiene were described under "personal safety regulations". The following hygienic provisions are available to comply with these rules.

10.5.1 Clothing
Every worker has a personal set of work-clothing, marked with his registration number and shift number. Every clothing packet consists of:
a white cotton drill overall with high neck-line, long trousers and long sleeves, a cotton drill washable cap, an interlock undershirt, a molleton overshirt, short interlock underpants, long interlock underpants, long molleton underpants, a pair of woollen socks, a bath towel, a neckcloth.
Together with this a cake of soap and a small pack of shampoo is supplied as well as a packet of paper handkerchiefs. Every insecticide worker has a pair of high leather shoes or leather boots free of charge, together with a pair of rubber slippers for use during lunch-break. Heavy grade seamless unlined PVC gloves are supplied for use in the insecticide plant and are being replaced by new ones whenever needed. Used gloves are destroyed.
None of the above mentioned materials may be taken home. Clothing has to be handed in every day at the laundry to be laundered. Shoes will be replaced and destroyed when worn out.

10.5.2 Washing facilities
Washing facilities consist of three partitioned spaces: a clean and a contaminated side separated by the shower-bath facilities. On the clean side there are lockers for private clothes and personal belongings: the clean work-clothes are issued here.
On the contaminated side every worker has another locker, where safety hat, plastic gloves, work-shoes and personal safety equipment are kept. From here everyone goes to his plant after placing his luncheon-packet in a special cupboard.
After work everyone goes in the opposite direction: used work-clothing is dumped into baskets on the contaminated side to be delivered to the laundry. Everyone takes the compulsory shower-bath and passes on to the clean side.

10.5.3 Lunchtime facilities

As eating is not allowed in the plants a special lunch-corner was created for insecticide workers. Different from their colleagues on shift work at other plants, insecticide workers have 20 minutes time off for lunch. First, shoes are put into the contaminated lockers, the rubber slippers are put on; after washing hands and face, the lunch-packet is taken from its cupboard and the worker passes on to the lunch-corner outside the insecticide plant. Here the overall is changed for a dust coat. On returning to work this same trip is made in the reverse.

10.5.4 The insecticide laundry

This laundry is specially created for laundering the work-clothes of the insecticide workers. Its equipment was described in chapter 9.2.4. It is operated on a 4-shift base as well, so that clothes can be laundered and ready again when the worker comes in for his next shift. In principle all clothes pass through the dry-cleaning machine and then get an after-treatment with water and soap.

Damaged or worn clothing are replaced here. Clean decontaminated work-clothes packets are assembled here and handed out. PVC-plastic gloves will be replaced in the laundry. Safety material has to be changed or handed in here, where it will be decontaminated before being returned to the safety material department for further cleaning and checking.

10.5.5 Miscellaneous

Thanks to good employee relations throughout the whole plant all workers feel completely free to visit the medical department at any time.

Good housekeeping is enforced and supervised.

Respiratory protective equipment is readily available everywhere in the plant.

Conveniently situated washbasins and safety showers are available throughout the insecticide plant.

10.6   SUMMARY

The different types of work and workers in the insecticide plant are described in this chapter, as well as the selection and the training of the workers, personal safety regulations and provisions for personal hygiene.

108

Chapter 11

Medical examination of exposed workers

11.1   INTRODUCTION

The industrial medical department of today has developed from a first-aid service, which was all we had up to 1952. In the beginning there was only a single first-aid man during daytime, and afterwards 4 men on shift duty. Medical supervision and medical assistance in case of accidents was given by the local general practitioner.

In April 1952 the first full-time company doctor started to work at this service, shortly thereafter followed by a second one, and from that time on development progressed over the years towards the industrial medical department of today with 4 full-time doctors qualified in industrial medicine and particularly acquainted with the specific toxicology, a first-aid centre with male nurses on shift duty, male industrial nurses, a clinical and biochemical laboratory, an X-ray department, a physiotherapeutical and revalidation centre and a medical administration. This development has not yet reached its final stage and in fact it must follow and where possible anticipate, new developments in medicine and technology.

When insecticide manufacture started at Pernis in 1954 this development was in its first phase. The following description of our medical examination schemes is a clear illustration of this progressive transformation. We will describe successively the development of the pre-employment examination, the pre-insecticide exposure examination, the routine medical examination of our exposed workers and the follow-up on those workers who were transferred to other plants on our site for one reason or another.

All medical interventions, such as workers not being accepted for insecticide work, preventive transfers for diverse reasons and transfers after intoxication with or without convulsions have been tabulated:
— in Table 11 according to calendar years,
— in Table 12 according to length of exposure.

These data refer to the total of 826 workers who have been employed full-time for a shorter of longer period in insecticide work between 1954 and January 1st 1968. Medical transfers as a result of insecticide exposure have been tabulated according to the insecticide causing the preventive transfer or intoxication. An extra group dieldrin + Telodrin had to be added for formulation plant workers who showed both insecticides in

Table 11

MEDICAL TRANSFERS, CHRONOLOGICALLY

Year	Total number of insecticide workers employed on July 1st of each year	Group 0: Applicants not admitted to insecticide work for medical reason	Group 1: Medical transfer without relation te exposure to insecticides	Group 2: Preventive transfer on account of suggestive non-incapacitating symptoms					Group 3: Clinical intoxications (all)					Group 4: Intoxications (convulsive ones only)				
				Diel-drin	Diel-drin/Telo-drin	Telo-drin	En-drin	To-tal	Diel-drin	Diel-drin/Telo-drin	Telo-drin	En-drin	To-tal	Diel-drin	Diel-drin/Telo-drin	Telo-drin	En-drin	To-tal
1954	40	–	–	–	–	–	–	–	–	–	–	–	–	–	–	–	–	–
1955	50	–	2	–	–	–	–	–	–	–	–	–	–	–	–	–	–	–
1956	80	–	2	–	–	–	–	–	2	–	–	–	2	2	–	–	–	2
1957	180	–	2	1	–	–	–	1	6	–	–	–	6	5	–	–	–	5
1958	220	–	1	1	–	–	–	1	6	–	–	2	8	4	–	–	2	6
1959	220	–	3	4	–	–	–	4	5	–	–	3	8	2	–	–	3	5
1960	220	–	2	–	–	–	–	–	–	–	–	–	–	–	–	–	–	–
1961	220	–	3	1	–	–	–	1	–	–	1	–	1	–	–	–	–	–
1962	230	2	7	1	–	–	–	1	1	3	3	1	8	1	1	3	1	6
1963	220	–	1	1	–	1	–	2	1	6	5	–	12	1	3	4	–	8
1964	210	11	15	3	3	16	–	22	2	–	5	–	7	–	–	1	–	1
1965	210	5	4	–	11	4	–	15	–	–	1	–	1	–	–	–	–	–
1966	200	3	2	1	4	–	–	5	–	–	–	1	1	–	–	–	1	1
1967	190	–	1	–	–	–	–	–	–	–	–	–	–	–	–	–	–	–
Total		21	45	13	18	21	–	52	23	9	15	7	54	15	4	8	7	34

Table 12

MEDICAL TRANSFER RELATED TO LENGTH OF INDIVIDUAL EXPOSURE

| | *Group 1* | *Group 2* | | | |
| | Medical transfer without relation to exposure to insecticides | Preventive transfer on account of suggestive non-incapacitating symptoms | | | |
Cumulative year of individual exposure in which intox. or medically advised transfer took place		*Dieldrin*	*Dieldrin / Telodrin*	*Telodrin* *Endrin*	*Total*
1st year of exposure	20	3	-	- -	3
2nd year of exposure	9	2	1	4 -	7
3rd year of exposure	4	6	5	5 -	16
4th year of exposure	3	1	2	2 -	5
5th year of exposure	2	-	3	2 -	5
6th year of exposure	1	-	-	1 -	1
7th year of exposure	2	-	1	1 -	2
8th year of exposure	3	-	1	4 -	5
9th year of exposure	1	1	-	1 -	2
10th year of exposure	-	-	3	1 -	4
11th year of exposure	-	-	1	- -	1
12th year of exposure	-	-	1	- -	1
Total	45	13	18	21 -	52

Group 3					Group 4				
Clinical intoxications (all)					Intoxications (convulsive ones only)				
Dieldrin	Dieldrin/Telodrin	Telodrin	Endrin	Total	Dieldrin	Dieldrin/Telodrin	Telodrin	Endrin	Total
10	-	1	5	16	8	-	1	5	14
8	2	2	2	14	5	1	1	2	9
5	2	5	-	12	2	1	1	-	4
-	1	1	-	2	-	-	1	-	1
-	1	1	-	2	-	-	1	-	1
-	1	2	-	3	-	1	2	-	3
-	1	2	-	3	-	-	1	-	1
-	1	-	-	1	-	1	-	-	1
-	-	1	-	1	-	-	-	-	-
-	-	-	-	-	-	-	-	-	-
-	-	-	-	-	-	-	-	-	-
-	-	-	-	-	-	-	-	-	-
23	9	15	7	54	15	4	8	7	34

their blood. As aldrin and isodrin are rapidly converted in the mammalian body into dieldrin and endrin respectively, intoxications due to these insecticides have not been mentioned separately, but have been incorporated in the dieldrin and endrin group respectively.

In Table 13 a specification is given of the medical reasons for those applicants or candidates who were not accepted for insecticide work and for those other workers who were later transferred from this plant for medical reasons other than those related to insecticide exposure.

Table 13

MEDICAL REASONS FOR REJECTION AND PREVENTIVE TRANSFER

Medical reasons for rejection or preventive transfer	Not accepted for insecticide work for medical reason (Group 0)	Preventive medical transfer without relation to insecticide exposure (Group 1)
Pre-existent EEG changes	11	12
Pre-existent skin diseases	2	8
Pre-existent mental disturbances: psychopathy, mental deficiency, senility, neurasthenia, etc.	3	8
Former or pre-existent neurologic diseases: epilepsy, migraine, meningitis, encephalitis, etc.	1	6
Former or pre-existent hepatitis, renal disease, hyperthyroidism or diabetes.	3	2
Pre-existent heart disease	-	2
Pre-existent pulmonary disease (CARA)	1	4
Stomach disorders, pre-existent.	-	3
Total	21	45

The insecticide intoxications mentioned in Table 11 and 12 will be discussed in detail in chapter 16.

11.2   PRE-EMPLOYMENT EXAMINATION

Pre-employment examination is done only for company-workers, not for contractor-workers. In some instances contractor-workers may have a pre-employment examination with their own company according to standards unknown to us.

As regards our pre-employment examination the following phases of development can be distinguished:

Until June 1952: pre-employment examination by outside contracted general practitioners. A short medical history was taken and general physical examination done. Simple urinalysis was done: tests for albumin and glucose only. A certificate of recent chest X-ray was requested. From June 1952-October 1953 this same procedure was continued by the company doctor.

October 1953-June 1959: a more extensive medical history and general physical examination by company doctors, otherwise the same. Since June 1959 a urine sediment was examined also, as well as a miniature chest X-ray done at our own site. In November 1960 a routine blood examination, consisting of Hb, SRE and WBC, was added as a routine, followed in June 1962 by blood group and rhesus factor.

In May 1969 we introduced a new general examination scheme to give maximum "base-line" information, against which future examination results can be compared. This new pre-employment examination scheme for every new company worker consists of:
— full medical history and general physical examination,
— miniature chest X-ray, audiometry, electrocardiography,
— vital capacity and forced expiratory volume (FEV),
— urinalysis: albumin, glucose, urobilin, urine sediment,
— blood examination: Hb, WBC, RBC, SRE, blood group, rhesus factor,
— a blood chemistry profile by Auto-Analyzer SMA 12/60 consisting of 12 biochemical tests: total protein, albumin, alkaline phosphatase, LDH, SGOT, inorganic phosphate, cholesterol, glucose, uric acid, bilirubin, calcium and urea nitrogen.

11.3   PRE-INSECTICIDE EXPOSURE EXAMINATION

After the first few years of insecticide manufacture, during which several acute insecticide intoxications occurred, it was felt that a pre-insecticide exposure examination should be required, particularly for the group of contractor-workers who did not have pre-employment examinations.

Due to various organisational problems this scheme could not be introduced before January 1962. From then onwards every new insecticide worker, company-worker as well as contractor-worker has a pre-placement examination and is not allowed to work in these plants before a written approval from the industrial medical department has been obtained. The examination procedure is similar to the pre-employment examination in the corresponding year, with only the following additions:
— baseline electroencephalography,
— baseline determination of insecticide levels in the blood since 1964,
— alkaline phosphatase and SGOT determinations since 1964, SGPT since 1967. A whole blood cholinesterase activity is determined in those workers who may also have contact with organophosphate insecticides.

From May 1969 onwards the full pre-employment examination as described in 11.2 is performed, extended with SGPT and of course determination of insecticides in blood.

This scheme applies to all future insecticide workers, operators as well as maintenance people, formulation hands, drum fillers and plant cleaners.

From Table 11 it can be seen that no rejections on medical advice are recorded prior to January 1962, the year in which our routine pre-exposure examination started. Since that time and up to January 1st 1968, 21 applicant or candidate workers were not admitted to insecticide work for various medical reasons, which are specified in Table 13. The peak of 11 rejections in 1964 is due to the rapid turnover of, in particular, contractor-workers in that year, which in its turn was caused by the relatively high number of preventive transfers in that year, which will be discussed later on. The fact that 1967 shows no rejections at all is an indication of the stability of the plant population in that year.

11.4 ROUTINE MEDICAL EXAMINATION OF EXPOSED WORKERS

From the beginning of 1955 onwards a routine periodic medical examination scheme for all insecticide workers — contractor- as well as company-workers — has been used. From 1955-1965 this examination took place twice yearly and included:
— medical and occupational history over the previous period,
— general physical examination with (once a year) miniature chest X-ray,
— urinalysis and blood counts, SRE, where necessary including differential counts,
— since 1956 electroencephalography is done as a routine method. This method was our principal tool for the detection of incipient insecticide intoxication up till 1964, when the following, more specific method was introduced,
— the determination of the levels of dieldrin, endrin and Telodrin in the blood was introduced as a routine in January 1964. From a preventive point of view this was a huge improvement over supervision by EEG control, since it provided an actual, though indirect, measurement of absorption for the individual as well as for the group. The preventive value of the periodic medical examination increased considerably as is obvious from the data given in chapter 13.

After having had experience with insecticide determinations in the blood for one year, we adapted our routine medical examination programme as from January 1st 1965 as follows:
— a general medical examination once a year only.
— EEG's once a year in aldrin-dieldrin production and formulation. We discontinued routine EEG's in endrin and intermediate production.
— determination of the insecticide levels in the blood once a year in intermediates and endrin production, and quarterly in aldrin-dieldrin production and formulation. The frequency of the blood level determinations to be adapted to circumstances: for instance during the later years of Telodrin production and formulation a monthly blood check was done on all exposed workers.
— to this routine scheme we added: alkaline phosphatase and SGOT in 1964 and SGPT in 1967. During 1967 SGPT, LDH, total serum proteins and serum protein-electrophoresis were done of all insecticide workers with insecticide exposures of longer than 4 years.

From May 1969 onwards all biochemical determinations have been replaced by a routine blood chemistry profile as discussed in 11.2 with additional SGPT determination.

These routine medical examinations of exposed workers provided results which led to two categories of transfers:
1. those for reasons not related to insecticide exposure,
2. those on account of suggestive non-incapacitating signs and symptoms from insecticide over-exposure.

These transfers will be discussed successively.

11.4.1 Medical transfers for reasons not related to insecticide exposure

Pre-exposure examination of insecticide workers only started in 1962, so that by that time quite a number of men — particularly in the group of contractor-workers — had slipped into insecticide work with a variety of former or pre-existent disease. For some workers of this group we advised transfer to other plants or other work when seen for their routine medical examination. Reasons for transfer are given in Table 13. These transfers may be regarded as delayed rejections for insecticide work (see 11.3); this is confirmed by the fact that, as can be seen from Table 12, nearly half of the transfers in this group took place during the first exposure year, i.e. at the first routine medical examination of those cases who had not had a pre-exposure examination.

11.4.2 Preventive transfers on account of suggestive non-incapacitating symptoms

During the earlier production years up to and including 1963 these transfers were made on account of suggestive but not incapacitating complaints or suggestive but not typical changes in the EEG pattern.

Since the introduction of the determination of insecticide concentrations in blood and the establishment of threshold levels for these insecticides, transfers in this group have been advised on account of these latter data only. Every worker with insecticide concentrations in his blood above the threshold level — all of them without symptoms or complaints — was transferred to other work. Some of them later returned to insecticide work, after insecticide levels in the blood had decreased considerably.

As threshold levels we took, as will be discussed in chapter 13:
— for dieldrin   : 0.20 μg/ml in blood,
— for Telodrin   : 0.015 μg/ml in blood,
— for dieldrin + Telodrin   : 0.20 D.Eq (see 2.3.3).

In Tables 11 and 12 these transfers have been tabulated in Group 2 and sub-divided according to the causative insecticide. From Table 11 it can be seen that Telodrin and Telodrin combined with dieldrin accounts for 75% of these transfers. From the same table it appears that the introduction of insecticide determination in blood in 1964 was the reason for 42 of all 52 transfers in this group. In 1967, 1968 and 1969 no more preventive transfers in this group were needed, a fact which demonstrates improvements in hygienic plant conditions, which in turn, indirectly to a large extent, resulted from the use of this test.

11.5   FOLLOW-UP ON FORMER INSECTICIDE WORKERS STILL UNDER SUPERVISION

Most former insecticide workers now working in other plants at out site are still under routine medical supervision. Those for whom an annual periodic medical examination is not routinely required have blood controls for insecticide levels, alkaline phosphatase, SGOT and SGPT, once a year. From May 1969 onwards also these people underwent an annual medical examination, including at least a blood chemistry profile, combined with determination of insecticide levels in the blood.

11.6 DISCUSSION

The results of this study of long-term industrial insecticide exposure are needed to answer the question whether long-term insecticide exposure at a dose which can be calculated, has any detectable effect in man, and whether from these data might be concluded that the current daily intake of the general population is a safe one.

To that end we have to examine first to what extent our industrially exposed population is a selected one, and thus, in how far the results of this study can be used to draw conclusions with regard to the general population (Zielhuis 1969).

From the above description of our medical examination schemes, one would at first sight gain the impression that a very stringent selection was made at different stages. However, in the preventive transfers that were not related to insecticide exposure (Group 1 in Tables 11, 12 and 13) there was admittedly no consistency in the way in which this was done. For example, one man who worried about potential ill effects of his work on a pre-existent disorder was transferred whereas his mate with the same disorder was not transferred. Some elderly contractor-workers which we felt not to be quite fit for this kind of work were left in their positions, at their own request, because they liked the work and the pay derived from it. Consequently, in Table 13 reasons will be given for those transferred, but as will be seen in chapters 14.1 and 16, quite a number of workers within the same diagnosis groups were allowed to continue at their work in insecticides.

We have no reason to believe that any one of the workers rejected or transferred would have come to harm in the work he was to do or was doing. Indeed we have many facts proving the contrary. In the case of pre-existent atypical EEG changes we have many workers who worked with insecticides for many years without mishap or alteration of their EEG pattern. This will be further discussed in chapter 16. Those workers with pre-existent disease, who were not rejected or transferred — as specified and discussed in chapter 14.1 — did not show any deterioration in the natural course of their disease while continuing work in insecticides.

In the decision whether to transfer or not, psychological reasons and financial consequences for these men themselves were sometimes considered of greater importance than the mere medical diagnoses.

The large number of 15 transfers in 1964 was caused by a specific reason; with the introduction of the technically more difficult handling of the more toxic Telodrin a re-appraisal of selection criteria became desirable. It was then decided to transfer all workers who were apprehensive, less disciplined or less skillful in order to avoid all possible future arguments of cause and effect relationships between work exposure and unrelated complaints or ailments. This change in selection and transfer policy was more for psychological and safety reasons rather than for pure medical considerations.

Of course workers with pre-existent skin diseases or with atypical pulmonary diseases form a "poor risk" in industrial handling of toxic material. This, however, should not be taken into account when industrial exposure data are used to assess the safety of exposure of members of the general population since the latter is practically totally confined to the oral route.

Up to this moment no medical transfers have been advised on account of workers receiving continuous medication and no signs of interference have been encountered.

As will be seen in chapter 16, transfers on account of insecticide intoxications and insecticide associated preventive transfers are related to hazardous situations as regards the specific properties of the insecticide. We never found that individual factors other than carelessness had been of influence in these cases, and thus caused a selection process.

11.7   SUMMARY

The development of the various medical examination schemes is described: pre-employment examination, pre-insecticide exposure examination, routine periodical examination of exposed workers, and follow-up examinations of former insecticide workers, as these schemes developed together with the development of our industrial medical department. This progressive development of examination schemes implies the drawback that base-line data from pre-employment examinations are not always available for all parameters, and if available not always fully comparable as other methods of determination have at times been used. From 1964 on, exposures as calculated from insecticide levels in blood can be compared with other examination results. From May 1969 on standardized biochemical profiles are available for future comparison.

All medical interventions in the group of 826 full-time insecticide workers up to January 1st 1964 have been tabulated chronologically per calender-year and related to the length of individual exposure. Reasons for rejection for, and transfer from, insecticide exposure were given. Transfers from insecticide work were classified according to the causative insecticide.

The selection policy in this group of 826 workers is discussed, from which the conclusion is derived that this group is comparable with any other working population.

D. RESULTS OF STUDY OF LONG-TERM OCCUPATIONAL EXPOSURE GROUP

Chapter 12

Composition and subdivision of long-term exposure group: general data and duration of exposure of subgroups

12.1 INTRODUCTION

Duration of exposure, together with insecticide blood levels at the steady state, form a good basis for comparison between workers industrially exposed to insecticides and the general population.

Of these two facets this chapter will deal with the duration of exposure and workers with long-term exposure will be assigned to appropriate groups. Blood levels of these long-term exposure groups will be discussed in the next chapter.

12.2 DIVISION INTO EXPOSURE GROUPS

Until January 1st 1968 a total of 826, only male, workers were employed in our insecticide plant on full-time jobs for a shorter or longer period. Maintenance workers, plant cleaners and formulators are included as well as operators.

Since we intended to study long-term industrial exposure, we had to make a selection from this total. Exposures shorter than one year (group A) were regarded as of little importance for drawing general conclusions. From workers with exposures exceeding one year, those with more than four years (group C) were selected arbitrarily to yield a sufficiently large group and with exposures long enough to evaluate the effects, if any, of relatively long-term occupational insecticide exposure. Thus our special attention in this part of the study will be centred on the long-term exposure group C.

This group C was subdivided into 3 subgroups:

C-3: those still working with insecticides;

C-2: those no longer working with insecticides but still under our medical supervision; and

C-1: those no longer working with insecticides and no longer under our supervision.

From the C-2 and C-3 groups it was possible to select three "extreme exposure" groups to serve as internal controls to these whole groups.

Thus, C-2-a is comprised of those workers transferred on medical advice because of high insecticide blood levels and C-2-b are the workers with former insecticide intoxication. Together they amount to about 1/3rd of the entire group C-2. Thus they constitute the high body-burden groups as compared to the total C-2 group.

C-3-a is a very long-term exposure group — each more than 10 years — forming about 1/3rd of the total C-3 group.

Thus degree as well as length of exposure are each provided with their own control groups, and we may assume that any trend or abnormality that would be present in the C-2 and C-3 groups would show up in a magnified form in one or more of these extreme exposure groups.

The subdivision into the various exposure groups will schematically be shown below and in Fig. 9.

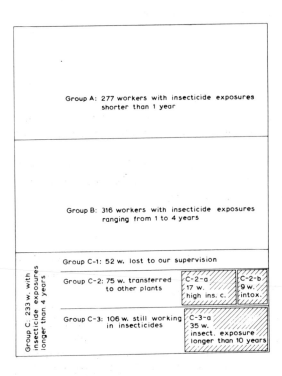

Fig. 9. Subdivision into groups of all 826 insecticide workers.

Subdivision of insecticide workers into exposure groups:

Group A:	277 workers with exposures shorter than 1 year.
Group B:	316 workers with exposures of 1-4 years inclusive.
Group C:	233 workers with exposures of between 4 years and 13¼ years as at January 1st 1968.
Total:	826 workers

Group C is subdivided as follows:

Group C-1:   52 workers who left the company and could not be contacted for supervision.

Group C-2:   75 workers who were transferred to other plants on our site but are still under our supervision.

Group C-3:   106 workers still working in our insecticide plant and having periodic medical examinations.

Total: 233 workers

Within the C-2 and C-3 groups the following "extreme exposure" groups can be distinguished:

Group C-2-a: 17 workers who were preventively transferred because insecticide levels in their blood exceeded intoxication threshold levels. The average time which elapsed after transfer was 2¾ years on January 1st 1968.

Group C-2-b: 9 workers who have had an insecticide intoxication. The average time which elapsed after intoxication was 6½ years on January 1st 1968.

Group C-3-a: 35 operators with an insecticide exposure of more than 10 years on January 1st 1968 and still working with insecticides. Average length of exposure in this group at that time was 11.1 years.

12.3 DURATION OF EXPOSURE AND OCCUPATIONAL STATUS

The whole period of insecticide work, including holidays, is included in the duration of exposure, which could be to the formulation, production of intermediates as well as to the production of the technical insecticides aldrin, dieldrin, endrin and Telodrin.

Because of the more or less frequent movement of workers between these units, a subdivision according to exposure to individual insecticides was not quite feasible; the less so as in our formulation unit all insecticides are handled at the same time or alternatively.

Ranges and arithmetic means of duration of exposure, as well as the total time of exposure of each C subgroup are given in Table 14; the distribution of exposure duration over the various subgroups is shown in Fig. 10. All exposures as at January 1st 1968.

From Table 14 and Fig. 10 can be seen that the average duration of exposure is similar in all groups with the exception of group C-3-a. This, however, was selected especially for extreme duration of exposure (more than 10 years). The other subgroups, we might conclude, are comparable as regards duration of insecticide exposure.

As parameters for occupational status we could compare:
a. the categories of skilled workers (operator, maintenance worker) with unskilled workers (formulation hand, drum filler, plant cleaner); and
b. the categories company-employed worker versus contractor-employed worker.

In most instances skilled workers are company-workers, there are, however, a few contractor-workers in maintenance work. Unskilled work is nearly always carried out by

124

Table 14
DURATION OF EXPOSURE

| Group | Number | Duration of exposure in years | | Group total in man·years |
		Range	Average	
C-1	52	4.0-12.3	6.6	341
C-2	75	4.0-12.3	7.1	532
C-3	106	4.0-13.3	8.4	895
C total	233	4.0-13.3	7.6	1768
C-2-a	17	4.0-11.6	7.7	131
C-2-b	9	4.7-12.3	7.7	69.5
C-3-a	35	10.0-13.3	11.1	400

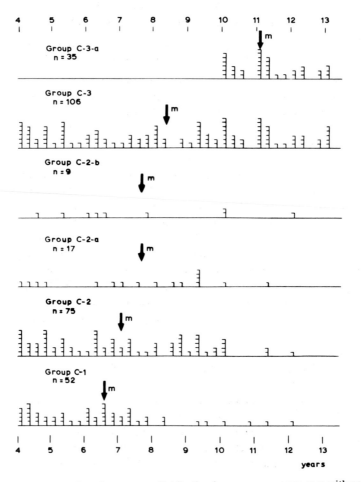

Fig. 10. Duration of exposure : distribution in exposure groups. m = arithmetic mean.

Table 15

COMPARISON OF OCCUPATIONAL STATUS OF EXPOSURE GROUPS

Group	Number	Skilled	Unskilled	Company	Contractor
A	227	128 = 46.2%	149 = 53.8%	129 = 46.6%	148 = 53.4%
B	316	220 = 69.6%	96 = 30.4%	227 = 71.8%	89 = 28.2%
C	233	192 = 82.4%	41 = 17.6%	185 = 79.4%	48 = 20.6%
C-1	52	42 = 80.8%	10 = 19.2%	38 = 73.1%	14 = 26.9%
C-2	75	70 = 93.3%	5 = 6.7%	68 = 90.7%	7 = 9.3%
C-3	106	80 = 75.5%	26 = 24.5%	79 = 74.5%	27 = 25.5%

contractor-workers, although some clean-out and drum filling workers are company-employed.

Skilled and/or company-employed workers in general have received more general and technical training, are more motivated towards their work and earn higher salaries than unskilled and/or contractor-workers. On average the latter category is less hygiene- and safety-minded than the former. Contractor-workers more readily move away to other work for various reasons; their "turnover" is thus greater than that of company-workers.

In Table 15 exposure groups are listed according to these criteria. From this it can be seen that especially group A (less than one year exposure), and to a lesser extent, group B (one to four year exposure), show high percentages of unskilled and/or contractor-workers when compared with group C. Thus, as far as occupational status is concerned, the A, B and C groups are not fully comparable.

Within the C subgroups the only striking difference is the much lower percentage of unskilled and/or contractor-workers in the C-2 group. This is explained by the fact that contractor-workers in the insecticide plant are not normally transferred to other plants for non-medical reasons, whereas this would not be unusual for company-employed operators and maintenance workers. When there are no specific reasons for transfer the contractor-workers either remain in the insecticide plant or the contractor moves them to other work outside the factory.

12.4   PERSONAL PARAMETERS

As personal parameters, age, body weight and height were chosen. Ranges and arithmetic means of these parameters per exposure group are listed in Table 16; in Figs. 11, 12 and 13 the distribution of these parameters per group is shown. Age per January 1st 1968. Weight and height as with the last medical examination prior to January 1st 1968.

For all these parameters variations in distribution, range and average are small. Therefore, the C subgroups are comparable as to age, body weight and height.

Table 16

PERSONAL PARAMETERS PER EXPOSURE GROUP

Group	Number		Personal Parameters		
			Age in yrs.	Body weight kg	Height cm
C-1	52	ar. mean	43.1	71.9	174.4
		range	25 - 68	50 - 89	159 - 198
C-2	75	ar. mean	40.4	74.4	176.3
		range	28 - 58	54 - 103	161 - 192
C-3	106	ar. mean	41.6	72.2	174.8
		range	22 - 64	55 - 98	157 - 190
C-2-a	17	ar. mean	39.1	75.8	174.9
		range	29 - 58	58 - 91	163 - 192
C-2-b	9	ar. mean	37.7	77.0	179.3
		range	30 - 51	64 - 92	166 - 190
C-3-a	35	ar. mean	42.8	70.7	174.7
		range	31 - 64	58 - 87	161 - 190

12.5 MEDICAL INTERVENTIONS

All transfers of workers on medical advice — see Tables 11 and 12, chapter 11 — are listed in Table 17 according to exposure groups. From this it can be seen that the 45 workers transferred for preventive reasons other than those related to insecticide exposure are fairly equally distributed between the groups. A higher percentage of such workers, however, belong to group A, which has relatively more contractor-workers who, during the earlier years of insecticide production, did not have a pre-employment examination.

The fact that transfer of insecticide workers because of high insecticide levels in the blood started only in 1964 accounts for the small number of these transfers in group A. The 3 preventive transfers in the C-3 group later returned to insecticide work.

Clinical intoxications are more or less equally distributed between exposure groups, as shown in Table 17.

The rate of convulsive intoxications to all clinical intoxications in the groups A, B and C is 12/13, 13/25 and 9/16 respectively. The high rate of convulsive intoxications in group A — 12 out of 13 clinical intoxications — is explained by the fact that in exposures of less than one year, many intoxications were due to acute gross over-exposure following

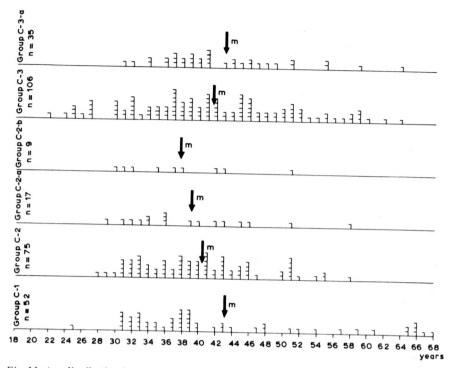

Fig. 11. Age distribution in exposure groups. m = arithmetic mean.

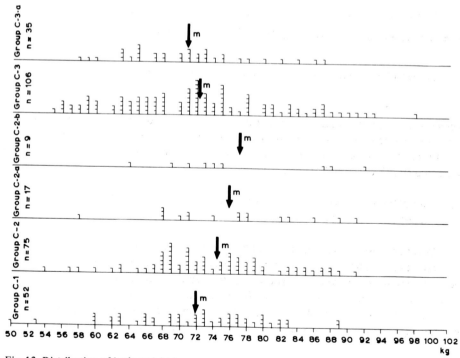

Fig. 12. Distribution of body weight in exposure groups.

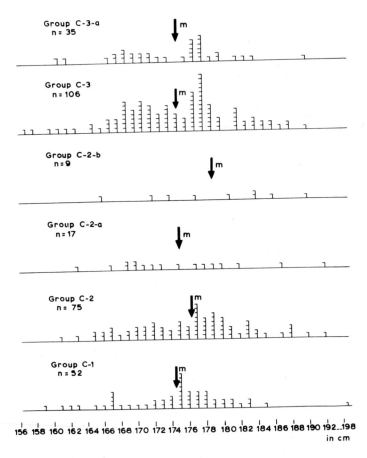

Fig. 13. Distribution of height in exposure groups. m = arithmetic mean.

accidents or careless handling. This again may be related to the higher number of unskilled contractor-workers in group A with jobs more exposed to insecticides.

From the foregoing we may conclude that the C group is not medically selected as compared with the A and B groups.

12.6 DELIMITING THIS PART OF THE STUDY AND ACCOUNTING FOR GROUPS NOT DISCUSSED

As already mentioned, the total of the workers studied and the division into groups, is based on the situation as it existed on January 1st 1968. Bringing this up to date would mean a total re-arrangement of groups and the figures derived therefrom: in reality a totally new study of exposure groups. Therefore this study of long-term exposure groups will include only data up to January 1st 1968. During the years 1968 and 1969 no facts or data were found which were in contradiction with those mentioned in this study.

However, following this study of long-term exposure groups, an up to date account of all insecticide workers will be given (chapters 17, 18, 19).

Since the object of our study is long-term insecticide exposure and its possible effects on the workers concerned, we shall confine our discussion of full medical details to the long-term exposure group C.

Full data concerning group A and B are available. These data contain nothing which is in contradiction with the data derived from the group C. Of these groups A and B highlights only will be discussed, *i.e.* cases of intoxication and transfers to other work for medical reasons.

Group C-1 contains 52 workers who left the company and are therefore no longer under our supervision and thus cannot be considered in full detail in this study. Here too, as with groups A and B, highlights only will be given.

Reasons for leaving the company in group C-1 were:

Shell employees:

Promotion elsewhere	10	
No promotional opportunities	4	
Starting own business	3	
Emigration	4	
Retired on a pension	1	
Own request: family reasons	6	
Objection to shift work	3	
Objection to work or environment	5	
Disciplinary	1	
Died of cancer of stomach	1	
	38	38

Contractor-workers:

Transferred to other work in employ of the same contractor	4	
Retired on a pension	5	
Unknown reasons	5	
	14	14
Total		52

We are certain that the reasons given for leaving employment with the company bear no relation to complaints or symptoms related to insecticide exposure.

12.7   SUMMARY

All 826 insecticide workers up to January 1st 1968 were divided into groups according to duration of exposure. A long-term exposure group (C) with exposures of more than 4 years was selected to study the effects on man, if any, of long-term exposure to the insecticides aldrin, dieldrin, endrin and Telodrin.

This long-term exposure group C was subdivided into 3 subgroups based on the work

Table 17
MEDICAL TRANSFERS RELATED TO EXPOSURE GROUPS

			Group 1	Group 2			
Exposure groups per 1-1-1968	Total number workers	Medical transfer without relation to exposure to insecticides	Preventive transfer on account of suggestive non-incapacitating symptoms				
			Dieldrin	Dieldrin/ Telodrin	Telodrin	Endrin	Total
Group A: Insecticide exposure <1 year	277	19	3	–	–	–	3
Group B: Insecticide exposure 1 - 4 years	316	16	7	7	8	–	22
Group C: Insecticide exposure >4 years	233	10	3	11	13	–	27
Total	826	45	13	18	21	–	52
Group C-1 who left the company and are lost to our supervision	52	5	2	3	3	–	8
Group C-2 who were transferred to other plants but are still under our supervision	75	5	–	7	9	–	16
Group C-3 still working in insecticide plant and having periodic medical examinations	106	–	1	1	1	–	3*

* Afterwards replaced in insecticides

situation as at January 1st 1968. From these subgroups extreme exposure groups were selected for intensity and duration of exposure respectively.

Groups A, B, and C proved to be different in occupational status. C subgroups are comparable in duration of exposure, age, body weight and height. Groups A, B, and C, as well as the C subgroups show no major differences due to medical selection.

Those 52 workers (C-1 group) no longer employed in our plants had reasons for leaving the company completely unrelated to signs or symptoms of intoxication or indisposition from insecticides.

The total long-term exposure group (C) consists of 233 workers with exposures ranging from 4-13.3 years, an average duration of exposure of 7.6 years and a total exposure of 1768 man-years.

Group 3					Group 4				
Clinical intoxications (all)					Intoxications (convulsive ones only)				
Dieldrin	Dieldrin/ Telodrin	Telodrin	Endrin	Total	Dieldrin	Dieldrin/ Telodrin	Telodrin	Endrin	Total
8	–	1	4	13	7	–	1	4	12
12	4	7	2	25	6	2	3	2	13
3	5	7	1	16	2	2	4	1	9
23	9	15	7	54	15	4	8	7	34
1	2	4	–	7	–	1	1	–	2
–	2	3	–	5	–	1	3	–	4
2	1	–	1	4*	2	–	–	1	3*

Chapter 13

Insecticide levels in the blood; half-life of insecticides in the body

13.1 METHODS

Heparinized whole blood samples were examined with the method described in chapter 2.3.2: first extracted with acetone, partitioned with hexane and then analyzed by gas chromatography using an electroncapture detector. Levels of the insecticides in the blood were determined in μg/ml.

Due to the rapid epoxidation of aldrin to dieldrin in the liver, only dieldrin needs to be determined in the blood and the aldrin-dieldrin exposure groups can be taken together.

Because many of our workers are exposed to different insecticides at the same time or subsequently, we use — as was explained in chapter 2.3.3 — for practical purposes, the Dieldrin Equivalent:

D.Eq = D + 10T, or

Dieldrin equivalent = dieldrin level + 10x Telodrin level. The D.Eq. in μg/ml is expressed as a dieldrin level.

In this study insecticide concentrations below detection levels were for statistical purposes considered to be one-half of that detection level.

Total insecticide levels of the various exposure groups resulting from combined exposures will be compared first. This will be followed by data on the individual insecticide levels. The half-life of insecticides in the body will be estimated from the available data.

13.2 INSECTICIDE LEVELS IN VARIOUS EXPOSURE GROUPS

Insecticide levels in the blood are the best available parameter of total insecticide intake and steady state body-burden.

Separate levels for the different insecticides, as well as the dieldrin-equivalent for each exposure group are listed in Table 18. Individual averages of all routine determinations made between January 1st 1964 and January 1st 1968 were calculated first. The means and ranges of these individual averages are then given for each group.

Table 18

MEANS AND RANGES OF INDIVIDUAL AVERAGES OF INSECTICIDE LEVELS IN BLOOD
BETWEEN 1.1.64 and 1.1.68 PER EXPOSURE GROUP

Group	No. of workers		Insecticide levels in blood in μg / ml			
			Dieldrin	Endrin	Telodrin	Dieldrin eq.
C-1 : former insect.	52	mean	0.056	n.d.*	0.0099	0.155
		range	<0.01-0.16		<0.002-0.030	0.02-0.40
C-2 : former insect.	75	mean	0.036	n.d.*	0.0105	0.141
		range	<0.01-0.12		<0.002-0.033	<0.01-0.38
C-3 : insect. workers	106	mean	0.025	n.d.*	0.0063	0.088
		range	<0.01-0.13		<0.002-0.024	<0.01-0.25
C-2-a : prevent. transfers	17	mean	0.045	n.d.*	0.0177	0.222
		range	<0.01-0.12		0.005-0.033	0.12-0.34
C-2-b : former insect. intoxication	9	mean	0.065	n.d.*	0.0225	0.290
		range	0.02-0.12		0.011-0.030	0.15-0.33
C-3-a : >10 years insect.	35	mean	0.029	n.d.*	0.0067	0.096
		range	<0.01-0.09		0.002-0.014	0.02-0.21

* n.d. = not detected at our detection level

Table 19

MEANS AND RANGES OF INSECTICIDE LEVELS IN THE BLOOD AT THE MOMENT OF
TRANSFER AND INTOXICATION RESPECTIVELY

Group	No. of workers		Insecticide levels in blood in μg/ml			
			Dieldrin	Endrin	Telodrin	Dieldrin eq.
C-2-a : prevent. transfers	17	mean	0.064	n.d.*	0.0222	0.286
		range	<0.01-0.16		0.010-0.041	0.21-0.43
C-2-b : former insect. intox.	9	mean	0.098	n.d.*	0.0233	0.331
		range	0.02-0.18		0.017-0.030	0.30-0.40

* n.d. = not detected at our detection level

Means and ranges of the blood levels in the extreme exposure groups C-2-a and C-2-b
at the moments of transfer and intoxication respectively, are listed in Table 19.

13.3 SUPPLEMENTARY DATA ON INDIVIDUAL INSECTICIDES

13.3.1 Dieldrin levels in blood

In fact we do not measure the dieldrin level in the blood, but the concentration of HEOD, the pure active ingredient which forms at least 85% of technical dieldrin. For simplicity's sake, however, we will speak of dieldrin levels.

In Table 20 dieldrin levels in the blood of our aldrin-dieldrin workers over the years 1964 up to and including 1969 are summarized, followed in the last column by the corresponding average equivalent oral daily intake of aldrin + dieldrin as calculated by the formula of Hunter and Robinson (1969; chapter 5.2).

Table 20
AVERAGE DIELDRIN LEVELS IN BLOOD PER YEAR IN ALDRIN-DIELDRIN WORKERS

| Year | Number of workers | Number of determinations | Dieldrin conc. in blood μg / ml | | Approximate average equivalent oral daily intake aldrin-dieldrin in μg |
			Mean (geometric)	Range	
1964	63	184	0.0690 (0.0609-0.0780)*	<0.01 - 0.44	802
1965	63	286	0.0597 (0.0550-0.0649)*	<0.01 - 0.24	694
1966	64	386	0.0497 (0.0464-0.0533)*	<0.01 - 0.22	578
1967	72	300	0.0310 (0.0290-0.0335)*	<0.01 - 0.18	360
1968	70	178	0.0320 (0.0294-0.0358)*	<0.01 - 0.22	372
1969	62	182	0.0247 (0.0217-0.0287)*	<0.005-0.15	287

* Confidence limits of means (P = 0.95)

Some of the workers might at that time not yet have reached the state of equilibrium, which means that their blood levels might still have been lower than would correspond to their average daily intake of insecticides. Thus the actual approximate average equivalent oral daily intake of aldrin + dieldrin in μg might well be somewhat higher than the values given, as calculated by the formula of Hunter and Robinson from the average blood levels.

From the data in Table 20 we may conclude that the insecticide exposure in our aldrin-dieldrin plant decreased with about 60 per cent over the years considered.

In our experience the threshold level below which signs or symptoms of intoxication do not occur, is at least 0.20 μg/ml in blood.

13.3.2 Endrin levels in blood

In 45 operators of our endrin plant blood concentrations have been determined at least once a year since 1964. We have never found endrin in the blood at our detection level of 0.01 μg/ml (since December 1965 0.005 μg/ml). Thus no worker exposed to endrin has ever been transferred on account of an elevated endrin level in blood.

On three occasions endrin could be detected in the blood of formulators handling endrin:

a. A drum filling hand was accidentally splashed with 20% endrin emulsifiable concentrate. He immediately took a shower bath with water and soap and changed clothes. A blood sample taken one hour after the accident contained: endrin 0.09 μg/ml; dieldrin and Telodrin below detection level. This man developed no signs or symptoms of intoxication. Five days later the endrin level in his blood was < 0.005 μg/ml.

b. A formulator handled technical endrin powder carelessly, producing a lot of dust and disobeyed instructions to wear a dust-mask. After having worked for 4 hours he sustained a convulsive seizure which was treated with phenobarbital 60 mg every 3 hours for one day. Afterwards he complained of headache only. The next day he felt perfectly well. The following blood levels were found:

Directly after the convulsive seizure:	0.08 μg/ml endrin,
	0.11 μg/ml dieldrin;
24 hours later:	0.02 μg/ml endrin,
	0.11 μg/ml dieldrin;
4 days after the 2nd sample:	<0.005 μg/ml endrin,
	0.10 μg/ml dieldrin.

Four colleagues of b. working next to him, but who had been wearing dust-masks were examined at the same time and showed endrin levels in the blood of 0.01, 0.01, 0.005 and <0.005 μg/ml respectively. None of these workers had any sign or symptom of intoxication, notwithstanding the fact that one of them had in addition to 0.01 μg/ml endrin, a dieldrin level of 0.18 μg/ml in the blood.

c. One of our operators was accidentally splashed with 20% endrin emulsifiable concentrate. He took a shower bath and changed clothes 10 minutes after the accident. He was kept under observation and prophylactically treated with phenobarbital: oral doses of 60 mg every 4 hours for 24 hours. He showed no sign or symptom of intoxication. The following blood levels were found:

40 min. after the accident: 0.027 μg/ml endrin, 0.01 μg/ml dieldrin;
12 hours after the accident: 0.025 μg/ml endrin, 0.01 μg/ml dieldrin;
36 hours after the accident: <0.005 μg/ml endrin, 0.01 μg/ml dieldrin.

We believe that his blood levels would have reached a higher peak between the first and the second sample i.e. between one and twelve hours after the accident probably 3 or 4 hours after accidental exposure.

In our opinion, which is in accordance with findings of others (chapter 7.1.1 and

7.3.2), endrin can only be detected in the blood, at our detection level, in situations of acute gross over-exposure..

The threshold level below which no signs or symptoms of intoxication occur, appears to be in the range of 0.050-0.100 μg/ml endrin in the blood.

13.3.3 Telodrin levels in blood

As Telodrin production ceased in September 1965 it is impossible to compare Telodrin levels in blood during successive production years as was done in the case of aldrin-dieldrin.

We have one case on record in which a Telodrin level in blood of 0.008 μg/ml due to acute over-exposure fell to below the detection level at <0.002 μg/ml within 3 days, probably due to re-distribution of the compound in the body.

In our experience the intoxication threshold level for Telodrin in the blood is 0.015 μg/ml.

13.4 HALF-LIFE OF INSECTICIDES IN THE BLOOD

As was discussed in the chapters on general toxicology of aldrin, dieldrin, endrin and Telodrin a certain daily dose of one of these insecticides causes a curvilinear (asymptotic) rise of the insecticide level in the blood and other tissues of the body until the state of equilibrium is reached between intake and excretion of the insecticide. When the administration of insecticide is discontinued, insecticide levels in blood and other tissues fall in the same asymptotic way until practically all insecticide has left the body. Thus we may consider the half-life (t½) of an insecticide in the body, which, since there is a fixed relationship between insecticide levels in the various body tissues, can be calculated from the half-life of an insecticide in the blood. A valid calculation is only possible in the state of equilibrium.

For dieldrin and Telodrin we selected groups of employees who were at a state of equilibrium as far as we could see from successive blood level determinations. For endrin this could not be done as endrin is not normally detected in the blood of our employees at our detection level. For completeness sake, aldrin should be mentioned here. Aldrin has a very short half-life in the blood because, within hours of entry into the body, it is epoxidized to dieldrin.

13.4.1 Half-life of dieldrin in the blood

For the purpose of estimating the half-life of dieldrin in the blood, we selected 15 aldrin-dieldrin workers with high dieldrin levels — at the state of equilibrium — in their blood, who were transferred to other plants for medical or non-medical reasons, and who could be followed-up. Half-yearly dieldrin levels in the blood were determined and tabulated, beginning with the half year in which transfer took place. Results are shown in Table 21.

From these data the half-life for each individual was calculated with its 95% confidence range. Combining the results of all 15 workers showed, that a biological half-life of 0.73 years fitted with all individual results (De Jonge 1970).

Table 21

DIELDRIN CONCENTRATION IN BLOOD IN $\mu g/ml$ AFTER DISCONTINUANCE OF EXPOSURE

Operator No.	Years since transfer					
	1/2	1	1 1/2	2	2 1/2	3
1	0.06	0.04	0.03	0.02	0.01	0.01
2	0.05	0.04	0.03	0.01	< 0.01	< 0.01
3	0.12	0.09	0.07	0.05	0.03	0.01
4	0.11	0.08	0.06	0.04	0.02	0.01
5	0.07	0.05	0.04	0.03	0.02	0.01
6	0.12	0.07	0.04	0.03	0.02	0.01
7	0.16	0.12	0.10	0.04	0.03	0.02
8	0.26	0.18	0.12	0.09	0.04	0.02
9	0.09	0.06	0.04	0.02	0.01	< 0.01
10	0.14	0.09	0.05	0.03	0.02	0.01
11	0.08	0.06	0.04	0.03	0.02	0.01
12	0.11	0.05	0.02	0.02	0.01	< 0.01
13	0.09	0.05	0.02	0.01	0.01	< 0.01
14	0.08	0.07	0.04	0.02	0.01	< 0.01
15	0.14	0.11	0.09	0.06	0.04	0.02

13.4.2 Half-life of endrin in the blood

As endrin is not normally detected in the blood of our employees at our detection level, only a few data concerning endrin levels in blood after acute over-exposure are known to us (see 13.3.2). As these peak exposures are quite different from a state of equilibrium in the body, the half-life of endrin cannot be estimated from the rapidity in which these endrin levels in the blood fall. However, from these data, which are in good agreement with data given in literature (chapter 7.1.1) and with results from animal experimentation, it is obvious that endrin is very quickly eliminated from the blood and from the body. The half-life of endrin in the blood may thus be of the order of 24 hours.

13.4.3 Half-life of Telodrin in the blood

As Telodrin production ceased in 1965, we compared Telodrin blood levels of 20 Telodrin workers at the state of equilibrium in the last production year 1965, with the levels in the four years following cessation of exposure to Telodrin. The results are shown in Table 22.

In the same way as for dieldrin (13.4.1) the biological half-life of Telodrin was calculated from these data to be 2.77 years (De Jonge 1970).

13.5 DISCUSSION

The mean dieldrin levels in the blood of our aldrin-dieldrin workers over the years from 1964 up to and including 1969 (Table 20), show a decline in the dieldrin blood level

Table 22

TELODRIN CONCENTRATION IN BLOOD IN $\mu g/ml$ AFTER DISCONTINUANCE OF EXPOSURE

Operator No.	Year				
	1965	1966	1967	1968	1969
1	0.016	0.012	0.009	0.007	0.006
2	0.012	0.010	0.007	0.005	0.004
3	0.021	0.016	0.013	0.009	0.006
4	0.024	0.016	0.015	0.014	0.013
5	0.025	0.015	0.010	0.005	0.006
6	0.026	0.024	0.014	0.010	0.008
7	0.018	0.015	0.009	0.008	0.007
8	0.012	0.009	0.006	0.005	0.004
9	0.022	0.023	0.012	0.011	0.007
10	0.043	0.020	0.019	0.014	0.015
11	0.036	0.032	0.016	0.018	0.014
12	0.027	0.018	0.012	0.013	0.007
13	0.012	0.010	0.007	0.007	0.006
14	0.021	0.014	0.010	0.008	0.007
15	0.010	0.010	0.005	0.003	0.003
16	0.008	0.007	0.005	0.003	0.002
17	0.072	0.056	0.040	0.039	0.031
18	0.007	0.006	0.006	0.005	0.003
19	0.008	0.008	0.006	0.004	0.003
20	0.020	0.015	0.012	0.013	0.007

from 0.0690 $\mu g/ml$ in 1964 (which is 115 times the present average blood level calculated for the general population in the U.S.A. and the U.K.: 0.0006 $\mu g/ml$) to 0.0247 $\mu g/ml$ in 1969 (which still is more than 41 times the blood level in the general population). If we look at the corresponding average equivalent oral daily intake of aldrin-dieldrin as calculated by the formula of Hunter *et al.* (1969), this also corresponds with a decline of the average daily intakes from 115 times to about 41 times that of the general population in the U.S.A. and the U.K.

This decrease in the intensity of insecticide exposure was the result of technical improvements and better hygienic plant conditions.

Although it was not possible to measure insecticide levels in blood in the years before 1963 other medical and technical observations would make it clear that the severity of insecticide exposure of our insecticide workers was at that time higher than in subsequent years.

Table 18, comparing insecticide levels of the long-term exposure groups, shows that average dieldrin levels over the four years up to January 1st 1968 fall in the same range, but that total insecticide exposures, expressed as the dieldrin equivalent, are much higher with blood levels ranging from 0.088 μg/ml D.Eq. for C-3 to 0.290 μg/ml D.Eq. for group C-2-b, with an average of 0.120 μg/ml D.Eq. for the total C-group. This is 143-483 times, and on an average 200 times, the D.Eq. level of the general population. If this were dieldrin only, we might apply the formula of Hunter *et al.*, and the exposures as expressed in average equivalent daily oral intakes aldrin-dieldrin, would range from 1023-3372 μg/man/day with an average of 1395 μg/man/day, corresponding to 146-482 times the average daily intake of the general population, with an average of 199 times this intake.

However, if we would disregard the Telodrin and endrin exposure, the average dieldrin blood level of the total C-group from January 1964 to January 1968 is 0.035 μg/ml, corresponding to an approximate average equivalent oral daily intake of aldrin-dieldrin of 407 μg/man/day, which is still 58 times that of the general population in the U.S.A. and the U.K.

As endrin has never been detected in the blood under normal working conditions at our detection level (0.005 μg/ml), we could not take into account daily exposures to endrin in our insecticide workers. However, we know from our daily industrial practice, that the exposure to endrin in the endrin plant is of the same order as that to aldrin-dieldrin in the aldrin-dieldrin plant.

The half-life as estimated in 13.4 of course is an average value, for, as was discussed in 2.1.3 there is a rather wide inter-individual variation in t½, as this is influenced by the many factors discussed there.

13.6   SUMMARY

The intensity of exposure in the long-term exposure groups — as selected (chapter 12) on account of the duration of their exposure — is discussed in this chapter.

If exposure to dieldrin alone is taken into account, the workers from these groups have an average dieldrin exposure of 58 times that of the general population in the U.S.A. and the U.K.

If the total insecticide exposure, as expressed in the dieldrin equivalent, is taken into account, then the average exposure was about 200 times as great as that of the general population. The C-2-a and C-2-b extreme exposure groups had still much higher average exposures.

The threshold levels of insecticides in the blood, below which we saw no sign or symptom of intoxication, are:

0.20 μg/ml for dieldrin

0.015 μg/ml for Telodrin and is believed to be:

0.050-0.100 μg/ml for endrin.

The half-life of insecticides in the blood, and thus in the body, estimated from our data are:

0.73 years for dieldrin,

about 24 hours for endrin and

2.77 years for Telodrin.

140

Chapter 14

General health of long-term exposure group

14.1 Medical histories
14.2 Routine medical examinations
    14.2.1 General physical examinations
    14.2.2 Chest X-rays
    14.2.3 Routine blood counts and urinalysis
14.3 Some pre- and post-exposure data of members of the extreme exposure groups, compared with those of a control group
14.4 Discussion of results
14.5 Summary

14.1 MEDICAL HISTORIES

Apart from the intoxications mentioned in chapter 12, Table 17, which will be further discussed in chapter 16, we found no diseases or accidents in the long-term exposure group, which could in any way be connected with exposure to insecticides. The absentee disease pattern in these groups is not different from that which we find in their non-insecticide-exposed colleagues. This disease pattern of all insecticide workers will be more fully discussed in chapter 17 on the pattern of absenteeism.

In chapter 11, Table 13 lists medical reasons for rejection and preventive transfer from the insecticide plant of workers with non-insecticide related diseases. In chapter 12, Table 17 these non-insecticide related transfers were divided as to subgroups. The necessity for these transfers might suggest that all workers who were more sensitive to insecticides or more prone to intoxication, were kept away from insecticide contact, and that thus the group of exposed workers would be selected in this respect. The reasons for these rejections and transfers were discussed in chapter 11.4.1 and may be compared to the listed pre-existent diseases in workers who were allowed to continue with insecticide work. Here we want to emphasize that those workers with pre-existent disease who were not rejected or transferred − with disease such as: hypertension (6), peptic ulcer anamnesis (5), asthmatic bronchitis (2), cardiac arrhytmias (2), arrested pulmonary tbc (2), rheumatoid arthritis (1), idiopathic thrombopenic purpura (1), psoriasis (1), relapsing dermatitis (1) − did not show any deterioration in the course of their disease while continuing to work with insecticides.

Sickness absenteeism in the group of our insecticide workers is not higher than among their non-exposed colleagues, as was shown by Hoogendam *et al.* (1965). Newer data on the pattern of absenteeism due to disease or accident will be discussed in chapter 17.

In the total long exposure C-group (n=233) no hepatic disease occurred at all. One case of infectious mononucleosis occurred, liver function tests of this patient were found normal after recovery.

The following diseases were recorded: one case of acute glomerulonephritis following furunculosis, 4 cases of nephrolithiasis, 3 cases of peptic ulcer, one case of mild diabetes, one case of hyperthyroidism, and one leg-amputation after accident. There were 2 cases of ischaemic heart disease; one patient of 47 was transferred after recovery to physically less strenuous work. The second patient, 46 years of age, died before returning to work. There was one case of extrasystoly in a 60 years old man.

As regards diseases of the nervous system: one case of lymphocytic meningitis occurred; one case of Bell's palsy fully recovering after surgical treatment, occurred 6 months after last contact with insecticides. One case of pressure neuritis of a foot occurred as a result of a swollen tendon.

In the neuropsychiatric field: one case of psychoneurotic kataplexy was observed. This patient improved after discontinuation of shift work. One worker suffered from pre-existent migrainous headaches.

In 3 workers of group C a malignant tumour was detected:
— a reticulosarcoma in a man of 44 years, whose father had died of leukemia.
— a carcinoma of the stomach in a man of 46 years who had had 5½ years of minor exposure working as a day foreman.
— a glioma in a worker of 51 years. This man had had frequent complaints of occipital headache for a long time prior to the onset of insecticide exposure. This very slow-growing tumour still allows him to continue working in another department.

14.2   ROUTINE MEDICAL EXAMINATIONS

As it was practically impossible to quantify all the results of the routine medical examinations, only aberrations from normal will be mentioned. Some of the parameters will be discussed in more detail in the next subchapter 14.3, where pre- and post-exposure values will be compared with those of a control group. Parameters used in liver profile will be discussed in detail in chapter 15.

14.2.1 General physical examination
Apart from signs and symptoms due to intercurrent diseases not related to insecticides, and the clear specific signs and symptoms from intoxication in some of the cases of overt poisoning, no abnormalities were detected in the long-term exposure groups.

We have never seen dermatitis of any type as a result of insecticide contact.

In the course of our routine medical examinations we found no increases in liver size as a result of insecticide contact.

14.2.2 Chest X-rays
In these long-term exposure groups no new pulmonary diseases were detected and no deterioration of pre-existing lung disease occurred.

14.2.3 Routine blood counts and urinalysis
In two cases a slight anaemia was detected in workers who had a previous gastrectomy. Four workers occasionally had a white count between 3500 and 4000 W.B.C.'s with a normal differential cell formula. Differential white blood cell counts were only done as a routine during the earlier production years and always proved to be

normal. Apart from pre-existent increased SRE's and those associated with the diseases mentioned above, 14 in total, one unexplained accelerated red bloodcell-sedimentation rate occurred. Naturally, in a few cases, *i.e.* in four workers, a few RBC's were occasionally found in the urine sediment without any indication of previous or concurrent renal disease. There were two cases of intermittent slight glycosuria. Although no statistical comparison was made, we believe this to be within normal limits for a group of 233 workers.

14.3 SOME PRE- AND POST-EXPOSURE DATA OF MEMBERS OF THE EXTREME EXPOSURE GROUPS, COMPARED WITH THOSE OF A CONTROL GROUP

In the extreme exposure groups the pre-exposure values of body weight, systolic and diastolic blood pressure, haemoglobin percentage (Hb), white bloodcell count (WBC) and the sedimentation rate of the red bloodcells (SRE) after one hour are compared with the values after a period of exposure to insecticides, the duration of which is given in chapter 12, Table 14. The insecticide concentrations in the blood of the people in these groups over the last four years are presented in chapter 13, Table 18. A similar comparison of data from a control group is presented.

For post-exposure data or data after more than 10 years exposure, the results of routine medical examinations or medical follow-ups performed between 1.7.1967 and 1.7.1968 were used on all occasions.

For controls we took a group of 25 operators of the HF-alkylation plant, which was started up a short time before the insecticide plant. These operators at that time also had routine medical examinations. For post-exposure data of this group, the results of a medical follow-up in September 1968 were used. There are no indications that the results in this group have been influenced by the exposures in this group.

Arithmetic means and ranges of all parameters are tabulated in Table 23.

Body weight was on all occasions determined under the same conditions. Although the intoxication group happened to have a higher initial body weight, all groups increased in weight over the years of observation, the controls and the workers who were transferred on account of high insecticide levels in blood perhaps somewhat more than the other groups.

As far as blood pressures are concerned, no further statistical methods are needed to show that long-term and/or intensive insecticide exposure does not increase systolic or diastolic blood pressure.

Hb concentrations before and after exposure are not fully comparable, for the determination series before exposure was performed by the Sicca method whereas the determination series after exposure by the cyano-Hb method, which has been used in our laboratory since April 1962. These methods have not been compared in our laboratory. It is obvious, however, that the trend in all groups is similar.

WBC's in all groups remain within normal limits with the variations that occur in normal healthy people. So do the SRE's.

Table 23

SOME PARAMETERS OF EXTREME EXPOSURE GROUPS COMPARED BEFORE AND AFTER EXPOSURE

			Group C-2-a preventive transfers bloodconc.	Group C-2-b insecticide intoxica-tions	Group C-3-a insecticide exposure >10 years	Control alkylation plant operators
Number of workers in group			17	9	35	25
Body weight in kg	before	average*	70.3	74.5	68.0	71.4
		range	50-78	61-94	58-86	53-96
	after	average	75.8	77.0	70.7	77.1
		range	58-91	64-92	59-87	57-100
Systol. blood pressure in mm Hg	before	average	134	137	133	130
		range	110-180	130-165	115-180	115-155
	after	average	139	129	132	136
		range	110-180	120-145	110-150	125-150
Diast. blood pressure in mm Hg	before	average	82	79	79	79
		range	60-90	65-90	65-90	55-95
	after	average	81	79	81	84
		range	60-90	70-90	70-90	80-95
Hb %	before	average	94.8	94.8	95.4	98.0
		range	83-103	88-103	84-110	85-110
	after	average	91.1	91.0	92.1	92.0
		range	83-100	82-95	81-103	82-105
WBC's	before	average	6300	6500	7300	6500
		range	4300-9200	4300-8900	4200-13500	4000-13700
	after	average	6600	5600	6300	5500
		range	4000-12000	3800-7200	3500-10100	3100-8900
SRE 1 hr.	before	average	3.8	2.8	4.3	3.6
		range	1-11	1-5	1-10(29)**	1-19
	after	average	5.9	3.5	6.2	6.4
		range	2-19	2-6	1-19(55)**	2-21

* Average is used in this table as short for arithmetic mean everywhere
** One case of rheumatoid arthritis

144

14.4   DISCUSSION OF RESULTS

A study of the medical files of workers with long-term insecticide exposure does not reveal findings which might be regarded as abnormal in groups of this size. In those workers with pre-existent diseases who continued work in the insecticide plant, no deterioration in the course of their disease occurred.

Workers transferred from insecticide work on account of either insecticide intoxication or insecticide levels in blood exceeding the treshold level, did not subsequently show an abnormal disease pattern during the period of observation.

In routine medical examination of all long-term exposure groups no unexpected abnormalities for a group of this size were found.

This part of the study is related to data up to January 1st 1968. Since then, more than two years have elapsed, and it is not possible to bring the tables up to date. It is, however, abundantly clear, that these two years longer continued exposure has in no way changed the favourable health condition presented and discussed in this chapter. This applies to the total exposure groups as well as to the selected extreme exposure groups, and to aldrin-dieldrin exposure as well as to endrin exposure.

A comparison of some parameters of extreme exposure groups either before and after exposure, or when compared to data from a control group (Table 23) does not show any trend which might be related to long-term or intensive insecticide exposure. The figures in Table 23 clearly point to the fact that long-term intensive insecticide exposure does not appear to induce hypertension and leucopenia in the high-exposure groups. That this would occur in the general population with very much lower exposures to these insecticides is therefore extremely unlikely.

14.5   SUMMARY

Medical files and routine medical examinations of the long-term exposure groups revealed no abnormalities other than those that would be expected in any group of 233 workers.

Body weight, blood pressure, Hb, WBC's and SRE's of the extreme exposure groups compared before and after 10 years exposure and compared to those of a control group did not show any effect from long-term intensive work with the insecticides aldrin, dieldrin, endrin and Telodrin.

Chapter 15

Parameters used in liver profile

15.1 Methods used
15.2 Results
15.3 Discussion of results
15.4 Summary

15.1 METHODS USED

Alkaline phosphatase, SGOT, SGPT and LDH were determined as well as total serum proteins and serum protein-spectrum.

Serum alkaline phosphatase was determined by the colorimetric method according to Bessey *et al.* (1946). The normal range for adults is considered to be 0.9-7.8 mMolU, with an average of 1.87 mMolU and a biological scatter (95% of normal observations) between 1.06 and 2.68 mMolU (1 mMolU = 16.7 mU/ml).

SGOT and SGPT were determined by the colorimetric method of Reitman and Frankel (1957) and measured in Karmen Units (K.U.; 1 K.U. = 0.482 mU/ml). Normal range for SGOT is 8-40 K.U. Normal range for SGPT is 5-35 K.U.

Serum LDH's were determined by an adapted method of Cabaud and Wroblewski (1958). Serum LDH activity is considered normal up to 300 mU/ml.

Total serum proteins and serum protein-spectrum were determined in the clinical laboratory of the Teaching Hospital "Dijkzicht", Rotterdam (Head Prof. Dr. B. Leynse). The serum protein-spectrum was determined by the electrophoretic method.

The normal ranges for the serum protein-fractions are included in Table 28.

15.2 RESULTS

Most parameters have been determined at more than one routine medical examination. Of some parameters not all samples were examined for the parameters concerned, or no results were received. All available observations have been included and as a result of the above the ratio of observations per worker is different in most groups.

In Tables 24-28 all results are tabulated according to the C subgroups. The number of determinations per group is given on all occasions. No data were available for the C-1 group. Range and average (arithmetic mean) are given for every group. In this series no 95% confidence ranges have been determined and no further statistical exercises have been done on account of above mentioned difference between groups in the ratio of observations per worker.

The results are tabulated as follows:

Table 24: Alkaline phosphatase
Table 25: S G O T
Table 26: S G P T
Table 27: L D H
Table 28: Total serum protein and serum protein-spectrum.

Table 24
ALKALINE PHOSPHATASE

Group	Number of workers	Number of samples	Alkaline Phosphatase in mMolU	
			Average	Range
C-2 : former insect.	75	76	1.70	0.9 - 3.4
C-3 : insect. workers	106	147	1.81	1.1 - 3.1
C-3-a : >10 yrs insect.	35	35	1.77	1.2 - 3.0
C-2-a : prevent. transfers	17	17	1.73	1.1 - 2.8
C-2-b : former insect. intox.	9	9	1.56	1.1 - 2.3

Table 25
SGOT

Group	Number of workers	Number of samples	SGOT in K.U.	
			Average	Range
C-2 : former insect.	75	75	17.40	5 - 45
C-3 : insect. workers	106	127	15.15	5 - 36
C-3-a : >10 yrs insect.	35	34	15.62	5 - 26
C-2-a : prevent. transfers	17	16	17.50	9 - 31
C-2-b : former insect. intox.	9	9	15.44	9 - 24

Table 26
SGPT

Group	Number of workers	Number of samples	SGPT in K.U.	
			Average	Range
C-2 : former insect.	75	65	17.23	6 - 34
C-3 : insect. workers	106	74	15.57	4 - 40
C-3-a : >10 yrs insect.	35	22	16.59	5 - 40
C-2-a : prevent. transfers	17	15	18.27	6 - 33
C-2-6 : former insect. intox.	9	8	11.75	6 - 16

Table 27
LDH

Group	Number of workers	Number of samples	LDH in mU/ml	
			Average	Range
C-2 : former insect.	75	71	163	83 - 350
C-3 : insect. workers	106	89	155	83 - 333
C-3-a : >10 yrs insect.	35	31	147	83 - 333
C-2-a : prevent. transfers	17	15	175	100 - 275
C-2-b : former insect. intox.	9	8	148	100 - 250

Table 28
TOTAL SERUM PROTEINS AND SERUM PROTEINSPECTRUM

Group	Number of workers	Number of samples		Serum proteins in g/l					
				Total	Albumin	Glob α_1	Glob α_2	Glob β	Glob γ
Normal Dijkzigt-Lab.			average	74.0	42.7	3.6	7.0	8.6	12.1
			range	62.0-86.0	34.1-51.3	2.2-5.0	3.7-10.3	4.9-12.3	6.3-17.9
C-2 : former insect.	75	71	average	71.36	40.75	2.65	7.30	7.64	12.89
			range	59.5-87.0	31.5-53.4	0.7-5.0	3.5-11.8	4.8-11.8	3.9-18.4
C-3 : insect. workers	106	89	average	68.99	39.36	2.82	7.04	7.30	12.41
			range	58.0-81.5	30.3-48.9	0.8-6.0	4.7-10.6	4.8-10.5	3.3-25.8
C-3-a : >10 yrs insect.	35	31	average	69.32	39.21	3.08	7.15	7.04	12.73
			range	62.0-78.0	31.3-47.8	1.7-4.8	4.7-10.1	4.8-9.2	5.4-25.8
C-2-a : prevent. transfers	17	15	average	71.37	40.47	2.81	6.56	7.55	13.57
			range	64.0-85.5	33.5-53.4	1.4-4.6	5.5-9.3	6.5-9.1	9.5-18.3
C-2-b : former insect. intox.	9	8	average	71.38	43.94	2.41	6.35	6.56	12.08
			range	64.5-78.0	35.6-51.4	1.4-3.2	4.6-7.8	4.8-9.8	6.5-16.0

148

15.3   DISCUSSION OF RESULTS

As far as acute effects are concerned, the no longer exposed C-2, C-2-a and C-2-b groups may be regarded as control groups to the still exposed C-3 and C-3-a groups.

The C-2-a and C-2-b, as well as the C-3-a subgroups are extreme exposure groups selected from the C-2 and the C-3 group respectively in the sense of high degree and long duration of exposure. Any effect of degree or duration of exposure should influence results of these subgroups when compared to the C-2 and C-3 group.

For alkaline phosphatase average and scatter are normal for all groups. 96% of observations fall within the biological scatter range.

Average and range of the SGOT's and SGPT's are normal. More than 98% of the results fall within the biological scatter range. There are no abnormally high results. Three very turbid serum samples in the C-2 group were not considered on account of unreliability.

Only one sample in groups C-2, C-3 and C-3-a exceeded 300 mU/ml for LDH.

For all enzymes there was no remarkable difference between C-2 and C-3 groups, or between C-2-a, C-2-b, C-3-a and the C-2 and C-3 groups.

Total serum proteins as well as the various components in all groups show normal averages and a normal biological scatter, although gamma globulins gave a few low results in the C-2 and C-3 group.

15.4   SUMMARY

Results of determination of alkaline phosphatase, SGOT, SGPT, LDH, total serum proteins, and serum protein-spectrum — tabulated per exposure group — do not show any indication of influence of degree or duration of exposure to these insecticides on these parameters.

E. RESULTS OF STUDY OF ALL INSECTICIDE WORKERS

Chapter 16

Patterns of intoxication – electroencephalograms – case histories

16.1 Introduction
16.2 Discussion of data on all cases of intoxication
16.3 Electroencephalograms
16.4 Patterns of intoxication
16.5 Case histories
16.6 Summary

16.1   INTRODUCTION

Since insecticide production at Pernis started in 1954, a total of 54 insecticide intoxications have occurred. These have already been mentioned in chapters 11 and 12 and tabulated:
– in Table 11 according to the calandar year in which they occurred,
– in Table 12 according to the length of individual exposure, and
– in Table 17 according to the exposure groups.
In all three tables all clinical intoxications are mentioned; cases with convulsive intoxications only are mentioned separately. In all tables intoxications have also been listed according to the causative insecticide: dieldrin (+ aldrin), endrin and Telodrin. For practical reasons, as was discussed in chapter 11.1, an additional group dieldrin + Telodrin had to be added.

In this chapter reference is made to Tables 11, 12 and 17. All other relevant data available in our medical files are also reviewed. The pattern of electroencephalographic tracings in organochlorine insecticide intoxication is discussed as well as the earlier and the current place of electroencephalography in routine medical examination of our insecticide workers.

The various types of intoxication by organochlorine insecticides as encountered among the 54 cases we have experienced, will then be discussed, followed by a few case histories of typical intoxication cases.

16.2   DISCUSSION OF DATA ON ALL CASES OF INTOXICATION

Hoogendam *et al.* (1962, 1965) described cases of intoxication which occurred during the earlier years of production. Those incidents were due partly to initial technical problems in the plants, partly to underestimating the risk of working with these toxic products and therefore to insufficient industrial and personal hygiene and inadequate selection of personnel. A description was given of improvements already made in these fields.

During the years 1962-1964 new cases of intoxication occurred, this time caused by technical problems in the Telodrin plant. Table 11 shows that 1964 was the last year in which an appreciable number of intoxications, mainly caused by Telodrin, occurred. As a result of technical and hygienic improvements and of the preventive transfer of workers with high insecticide levels in the blood, we had fewer problems after that time. The last Telodrin intoxication occurred in January 1965, and one acute endrin intoxication due to acute gross over-exposure occurred in 1966. Since then no insecticide intoxications have occurred in our plant, despite increased production and formulation of technical insecticide with less man-power.

Some data on personal measurements of the workers involved in intoxication are tabulated in Table 29, grouped according to the exposure groups described in chapter 12. This table shows that the average age at the time of intoxication, 33.8 years, is lower than the average age of our insecticide workers: slightly more than 40 years at January 1st 1968 (group C in Fig. 11).

Table 29

DATA ON ALL CASES OF INSECTICIDE INTOXICATION

		Group A: <1 year in insecticides	Group B: 1-4 years in insecticides	Group C: >4 years in insecticides	All groups
Number of intox.		13	25	16	54
Exposure length	Arithm. mean	0.66 yr	2.04 yr	5.23 yr	2.62 yr
	range	0.01-0.9 yr	1.0-3.9 yr	4.0-7.7 yr	0.01-7.7 yr
Age in years	Arithm. mean	34.6	34.2	32.4	33.8
	range	19-53	24-63	26-42	19-63
Weight in kg	Arithm. mean	68.4	72.3	73.5	71.9
	range	59-85	59-95	60-92	59-95
Length in cm	Arithm. mean	172.8	174.5	176.2	175.0
	range	162-184	164-188	166-190	162-190
Unskilled workers		11=84.6%	9=36%	4 = 25%	24 = 44.4%

The average body weight does not differ from the average weight of the C group (Fig. 12), and the average body length is in the same range as the average length of the C group (Fig. 13).

From Table 29, as compared with Table 15, it is clear that intoxications occurred relatively more often associated with unskilled work such as drum filling, etc. As is shown in Table 30 this is particularly true for aldrin-dieldrin work. Only in Telodrin work did trained operators show an appreciable number of intoxications.

From the data in Table 12 the ratio of convulsive intoxications to total intoxications in the case of the various insecticides and as regards the cumulative individual exposure years can be calculated. The results are shown in Table 31.

In the case of endrin, all 7 intoxications took place in the first two years of exposure and all were accompanied by a convulsive seizure. Looking at the medical histories of

Table 30

INSECTICIDE INTOXICATIONS DIVIDED AS TO CAUSATIVE INSECTICIDE AND SKILLED OR UNSKILLED WORK

Causative insecticide	Unskilled workers	Skilled workers	All workers
Aldrin-dieldrin	15	8	23
Endrin	3	4	7
Telodrin	2	13	15
Dieldrin + Telodrin	4	5	9
All insecticides	24	30	54

Table 31

RATE OF CONVULSIVE INTOXICATIONS IN TOTAL INTOXICATIONS FOR THE VARIOUS INSECTICIDES PER INDIVIDUAL EXPOSURE YEAR

Exposure year	Endrin	Dieldrin	Telodrin and dieldrin + Telodrin
1st	5 in 5 = 100%	8 in 10 = 80%	1 in 1 = 100%
2nd	2 in 2 = 100%	5 in 8 = 62½%	2 in 4 = 50%
3rd	–	2 in 5 = 40%	2 in 7 = 28½%
4th	–	–	1 in 2 = 50%
5th	–	–	1 in 2 = 50%
6th	–	–	3 in 3 = 100%
7th	–	–	1 in 3 = 33½%
8th	–	–	1 in 1 = 100%
9th	–	–	0 in 1 = 0%
Total	7 in 7 = 100%	15 in 23 = 65%	12 in 24 = 50%

these 7 men, one finds that all 7 endrin intoxications were due to acute gross over-exposure either following an accident or due to obvious neglect of precautions. In all 7 cases the recovery to normal was rapid: within 1-3 days.

In the case of aldrin and dieldrin all intoxications occurred, for each individual worker afflicted, in the first 3 years of exposure. Intoxications during the first year of exposure followed the same pattern as described above for endrin, with convulsive seizures in nearly all cases. In the medical files only one case of aldrin-dieldrin intoxication following acute gross over-exposure was found. In the 2nd and 3rd year of exposure the pattern of intoxication changed in that not all intoxications were peracute but cases appeared in which either the convulsion was preceded by prodomi, or a gradual increase in concentration of dieldrin in the blood occurred together with prodomi but without culminating into a convulsion.

As can be seen in Table 30 the majority of cases of aldrin-dieldrin intoxication occurred among unskilled workers. This was probably due to insufficient plant hygiene at the filling points where technical and formulated insecticides were handled and where most unskilled workers were employed. After plant hygiene and technical filling equipment were considerably improved problems with aldrin and dieldrin no longer occurred.

In all cases of aldrin-dieldrin intoxication recovery to normal was rapid. Occasionally a subsequent convulsive seizure occurred on the first or second day, followed by minor subjective complaints for up to two weeks. No occurrence of a convulsion was observed more than two days after cessation of exposure to aldrin, dieldrin or endrin.

In our experience patients with a peracute convulsive intoxication, due to an obvious or known acute over-exposure, recover more promptly than workers in whom the convulsive intoxication followed a more gradual accumulation of the toxicant in the body due to a more prolonged less massive over-exposure.

In the Telodrin group of workers as well as in the dieldrin + Telodrin group the picture is quite different. When blood analysis became available it appeared that comparatively high Telodrin levels in the blood were more persistent, both in workers who continued and in those who were removed from exposure. Even in cases with intoxication by Telodrin without convulsive seizures there were many complaints about pronounced headache (sometimes with "snapping" sensations in the head), dizziness, drowsiness and pronounced irritability, sometimes with paresthesia's, particularly in the legs.

Although full recovery to normal health occurred in all patients with intoxications, in the case of Telodrin this took longer than with the other insecticides. Thus in three of our cases some typical complaints persisted for six months. Also the return to normal of

Table 32

INSECTICIDE LEVELS IN THE BLOOD AT THE TIME OF INTOXICATION AND AT THE MOMENT OF TRANSFER ON ACCOUNT OF INSECTICIDE LEVELS EXCEEDING SAFETY THRESHOLDS

Group	No. of workers		Insecticide levels in the blood in μg/ml			Dieldr. Equiv.
			Dieldrin	Endrin	Telodrin	
C-2-b:						
insecticide intox.	9	mean	0.098		0.0233	0.331
		range	0.02-0.18	<0.005	0.017-0.030	0.30-0.40
All insecticide intox.	20	mean	0.054		0.0158	0.320
		range	<0.01-0.24	<0.005	0.004-0.060	0.22-0.69
C-2-a:						
preventive transfers	17	mean	0.064		0.0222	0.286
		range	<0.01-0.16	<0.005	0.010-0.041	0.21-0.43
All preventive transf.	40	mean	0.055		0.0184	0.303
		range	<0.01-0.22	<0.005	0.002-0.051	0.21-0.53

the changed EEG pattern took longer than in the case of the other insecticides, sometimes more than a year.

This difference can easily be explained if one considers that the half-life of Telodrin in the blood is approximately 2¾ years. This long biological half-life of Telodrin may well have contributed to the higher incidence of Telodrin intoxications in skilled operators (Table 30).

In 8 out of 24 intoxications due to Telodrin (+ dieldrin) an acute over-exposure may have played an additional role.

In Table 32 insecticide levels in the blood are given of all the cases in which they were known at the time of intoxication or preventive transfer.

For comparison, corresponding data of the extreme exposure groups C-2-a and C-2-b, (already mentioned in Table 19) are repeated here. These results are fully comparable and show that mean D.Eq. at the time of preventive transfer and intoxication were 0.303 and 0.320 μg/ml respectively, or 505 and 533 times the present D.Eq. of the general population in the U.K. and the U.S.A.

16.3   ELECTROENCEPHALOGRAMS

In chapter 2.3.1 the use of electroencephalograms as a parameter for indicating imminent convulsive intoxication with organochlorine insecticides was discussed. For details we refer to this chapter.

The two papers by Hoogendam et al. (1962, 1965) dealt mainly with electroencephalography in our insecticide workers. This will not be repeated in detail. Up to the present moment more than 2500 EEG's of our insecticide workers have been taken and interpreted by M. de Vlieger M.D. (Department of Neurology, Teaching Hospital "Dijkzigt" Rotterdam).

As already mentioned, routine electroencephalography as a preventive method has almost entirely been replaced by determination of insecticide levels in the blood. We still use electroencephalography as a pre-exposure screening method, in cases of suspected intoxication, and as a yearly routine in aldrin-dieldrin workers.

From the time that exposures were discontinued at or before threshold levels of insecticide concentrations in the blood were reached, typical EEG changes have no longer occurred. Below this threshold level there is no correlation between EEG pattern and insecticide concentration in the blood. Sometimes non-typical EEG changes are seen with low insecticide concentrations in the blood. These probably represent the wide scatter in normal EEG patterns.

In all cases of insecticide intoxication associated with typical EEG changes which could be followed-up, EEG patterns returned to normal. In the case of pre-existent atypical EEG changes, they returned to their former pattern. This is illustrated in Table 33, where results of electroencephalography in the extreme exposure groups are tabulated: those of operators who worked for more than 10 years in insecticides (C-3-a), those of workers transferred on account of blood levels exceeding our safety threshold levels (C-2-a) and those of workers transferred on account of insecticide intoxication (C-2-b). The EEG results of all cases of intoxication cannot be tabulated as some of the pre-exposure data and some of the follow-ups are lacking. All available data, however, are

Table 33

ELECTROENCEPHALOGRAMS IN EXTREME EXPOSURE GROUPS

Group	Number of workers	Number of EEG's		Non-intoxications EEG's			EEG's typical for intoxic.		
		Total	Average per man	Always normal	Some-times atypical changes	Always atypical changes	Normal before and after intox.	Suspect changes on one occasion befo after into.	Atyp char befo after into.
C-3-a : >10 yrs insect.	35	265	8	22	8	3	2	–	–
C-2-a : prev. transfers	17	133	8	11	3	2	–	1	–
C-2-b : insect. intox.	9	82	9	–	–	–	7	–	2

consistent with the description given of the results of the extreme exposure groups.

Another important point is, that in cases with pre-existent atypical EEG changes, intoxications occurred at insecticide concentrations in blood approximately equal to those of workers with a previously normal EEG pattern. Thus it appears that people with pre-existing atypical EEG changes are not more susceptible to intoxication. No signs of intoxication were found in EEG's of workers with insecticide levels in blood below threshold levels.

The precise time it took for EEG's to return to the pre-exposure pattern after intoxication is difficult to assess from our data since follow-up EEG's were made after rather long intervals. So if for instance we found a normal pattern half a year after intoxication, this might well have already been normal after one or two months. Taking into account all the information we have, it is estimated that if insecticide exposure were discontinued after intoxication, the return to the original EEG pattern took never longer than at most a few months in endrin, at most a year in dieldrin, but sometimes more than a year in Telodrin intoxications. However, in many cases recovery was much quicker.

16.4   PATTERNS OF INTOXICATION

It has already been said (i.e. in chapter 2.4.1 in the introduction to insecticide intoxications and in the chapters on the toxicology of the insecticides in man) that as regards symptomatology, it is not possible to distinguish between causative insecticides on the basis of signs and symptoms only.

However, these signs and symptoms may occur in different patterns as a result of different types of intoxication, depending on the rate of exposure, the acute toxicity and the persistence in the body of the insecticides.

In 2.4.1 Hayes' (1963) classification of dieldrin intoxication into various types was quoted.

Hayes' 3rd type in which "many relatively small doses may produce one or a few convulsions with lesser accompanying symptoms that may recur even though exposure is discontinued", is an intoxication type which we have not encountered in any of our 54 cases of intoxication. Looking at the pharmacokinetics of the insecticides under review, we believe that this type of intoxication is not likely to occur frequently.

Based on our experience we suggest that Hayes' types of intoxication be slightly modified as follows:

Type 1: an acute convulsive intoxication with no or only minor prodomi resulting from one or a few gross over-exposures. This type tends to occur particularly with highly toxic and not very persistent insecticides. All our endrin intoxications as well as the endrin food poisonings described in the medical literature belong in this category. Occasionally cases of intoxication by aldrin and dieldrin also follow this pattern.

Type 2: a greater number of smaller doses of an insecticide with a lower acute toxicity but a higher persistence may cause an accumulative intoxication. Clinically, this results in a syndrome of headache, dizziness, drowsiness, hyperirritability, general malaise, nausea, anorexia, occasional vomiting. At times muscle twitchings, myoclonic jerks and convulsions occur. In these circumstances minor increases in the insecticide level in the blood, perhaps caused by minor fluctuations in exposure, may bring about a convulsive intoxication. This type of intoxication was seen in many cases caused by aldrin, dieldrin and Telodrin.

Type 3: this is actually a combination of type 1 and 2 and this type was also encountered in our aldrin, dieldrin and Telodrin workers. In this type an over-exposure, in itself not significant, causes an acute convulsive intoxication superimposed upon a subclinical accumulative intoxication of type 2.

These 3 types of intoxication are schematically illustrated in Fig. 14.

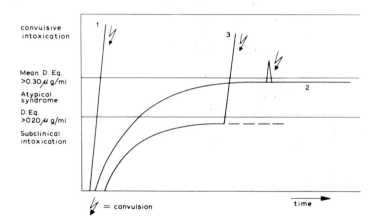

Fig. 14. Three types of organochlorine insecticide intoxication

16.5   CASE HISTORIES

In chapter 13.3.2 three typical cases of gross over-exposure to endrin were described.

Hoogendam *et al.* (1962) described a case of convulsive intoxication following accidental gross over-exposure to isodrin in 1957. This operator is still working in our endrin plant. In his medical history since that time nothing is found which might in any way be related to this over-exposure.

The following case histories do not include all medical details which we have on record; only those headlines which are important to illustrate cause and pattern of intoxication, as well as recovery, are mentioned.

Case history 1

A 43 year old operator was grossly over-exposed to endrin while carelessly cleaning a clogged centrifuge during his morning shift. In the afternoon without prodomi epileptiform convulsions occurred from which he fully recovered within 3 days. Medical examination one day after the convulsions did not reveal any abnormality. Insecticide levels in the blood could not be determined at that time. Two days after the convulsions there were slight changes in the EEG, which on re-examination one month later had returned to normal.

Case history 2

A construction painter aged 42 years was scraping off and painting equipment in the endrin plant (1959). He wore his own clothes during work and did not take a shower-bath afterwards. In the evening after the second day of work he was hospitalized with a convulsive intoxication. An EEG made a few days afterwards showed some dubious changes. Recovery was uneventful.

Case history 3

In 1958 after having worked for 10 months in drum filling of formulated aldrin-dieldrin products, a 34 year old contractor-worker complained of headache, drowsiness and dizziness. He was nervous, had paraesthesias in arms and legs and myoclonias in his jaw muscles. An EEG showed typical changes, whereas previously it had been normal. After transfer to other work, signs and symptoms disappeared within two weeks, and 3 months afterwards the EEG had returned to normal.

Case history 4

A 29 year old formulation hand and drum filler was regularly over-exposed to the various insecticides due to technical imperfections of the filling equipment. He had done this work already for more than 5 years, when in January 1964, while working on a rush order of Telodrin 50% wettable powder, he had an epileptiform convulsion. General examination and laboratory tests immediately afterwards did not reveal any abnormalities.

Insecticide levels in the blood were:

dieldrin 0.18 μg/ml,

endrin not detected,

Telodrin 0.022 μg/ml.

Typical EEG changes were found which had reversed after one month. He was sent home and felt perfectly healthy one week afterwards. His further medical history is uneventful.

Case history 5

A 24 year old operator who had been working in the formulation unit for 1½ years had an epileptiform convulsion in March 1963 with typical EEG changes. He had no signs or symptoms before or afterwards. The EEG pattern was also normal before and afterwards.

Insecticide levels in the blood at the moment of intoxication were:
dieldrin 0.15 μg/ml,
endrin not detected,
Telodrin 0.009 μg/ml.

Case history 6

A 29 year old operator after one year aldrin-dieldrin work complained in 1957 of increasing headache, dizziness and paresthesias in his fingertips. Typical EEG changes were found. He was transferred to another plant. Signs and symptoms disappeared within two weeks. Re-examination of the EEG 6 months later showed a normal pattern again.

Case history 7

A 36 year old unskilled worker had been working for more than a year in the formulation unit. He felt perfectly healthy and showed no abnormalities on routine medical examination in April 1964. However, insecticide levels in the blood were:
0.23 μg/ml dieldrin,
endrin not detected,
0.011 μg/ml Telodrin.
On taking the EEG he had myoclonias during hyperventilation and fotostimulation. He was transferred to other work. Three months later there were no abnormalities in his EEG.

Case history 8

A 39 year old operator had been working for 4 years in aldrin-dieldrin and also 2 years in Telodrin manufacture. Already in the aldrin-dieldrin period in 1957 and 1959 he had periods of headache, dizziness and drowsiness. His EEG was slightly abnormal on one occasion in 1959. Some time after transfer to the Telodrin unit his complaints increased: headache, lassitude, drowsiness and paresthesias in the legs. In August 1963 he had epileptiform convulsions at home followed by myoclonias and black-outs. The EEG then showed the typical pattern of organochlorine insecticide intoxication. Insecticide levels in the blood at that time were:
dieldrin 0.022 μg/ml,
endrin not detected,
Telodrin 0.028 μg/ml.
He was transferred to other work but had atypical complaints for about 6 months afterwards. Then the EEG had improved considerably and blood levels were:
dieldrin 0.01 μg/ml,
endrin not detected,
Telodrin 0.016 μg/ml.
The EEG pattern has returned to normal since then.

Case history 9

A 36 year old operator had worked uneventfully for 5 years in aldrin-dieldrin manufacture. One and a half years after transfer to the formulation unit he had a black-out without any other signs or symptoms. Insecticide levels in the blood were then:
dieldrin 0.13 μg/ml,
endrin not detected,
Telodrin 0.011 μg/ml.
There were no EEG changes at any time.

Case history 10

A 34 year old operator had worked uneventfully in aldrin-dieldrin and Telodrin manufacture for 4 years. He had never had any signs or symptoms of intoxication and his EEG's always showed a normal pattern. In July 1963, while walking with one of his children he had a black-out. Next day typical EEG changes were found and the insecticide levels in the blood were:

dieldrin 0.020 μg/ml,

endrin not detected,

Telodrin 0.030 μg/ml.

He was transferred to other work and had no more signs or symptoms of intoxication and since that time no diseases or accidents worth mentioning. The EEG was repeated 2 months after the incident when it had returned to normal.

Case history 11

A 30 year old operator had worked in endrin for 4 years when he was transferred to Telodrin manufacture where he worked for 9 months. In January 1962, five hours after a gross accidental over-exposure to Telodrin, he had epileptiform convulsions at home, fell against a stove and burned his cheek and nose. He had backache for quite some time afterwards. The EEG showed typical changes, which 3 months later had returned to normal. Blood levels of insecticides could not be determined at that time, but 2½ years later we found:

dieldrin and endrin <0.01 μg/ml,

Telodrin 0.006 μg/ml.

This operator was transferred to other work, but had atypical complaints for 6 months after the accident. The further course was uneventful.

Case history 12

A 26 year old operator had been working uneventfully for 4 years in aldrin-dieldrin and Telodrin manufacture. During the last year he had been working on the centrifuge of the technical Telodrin, work which was associated with high exposure. In May 1963 he complained about headache, "snapping" sensations in the head and myoclonias. The EEG, which previously had been normal, showed typical changes. Insecticide levels in the blood were:

dieldrin 0.03 μg/ml,

endrin not detected,

Telodrin 0.030 μg/ml.

He was transferred to insecticide intermediates where insecticide exposures are lower, but two months afterwards, after swimming during the afternoon, had an epileptiform convulsion. The EEG was then still abnormal and insecticide levels in the blood were:

dieldrin 0.06 μg/ml,

endrin not detected,

Telodrin 0.026 μg/ml.

He was transferred to another plant. Atypical complaints and at times "snapping" sensations in the head continued for about 6 months. An EEG, repeated one year later showed a normal pattern. The further course was uneventful.

16.6   SUMMARY

From an examination of medical data of all 54 insecticide intoxications which occurred in our organochlorine insecticide plant, it became evident that most intoxications had occurred in the group of unskilled workers employed in jobs in which insecticide exposure was higher than for most trained operators. The only exception was in the case of Telodrin exposure where, probably due to its persistence in the body, several operators had intoxications.

The average age of the group which suffered intoxications was below that of all insecticide workers. Body weight and length of the intoxication group did not differ from the average of all insecticide workers.

All endrin intoxications and one aldrin-dieldrin intoxication were due to accidental gross over-exposure. In all other cases of intoxication there was a more or less accumulative type of intoxication, sometimes with superimposed over-exposures.

No signs or symptoms of intoxication were found when concentrations of insecticides in the blood were below our safety threshold levels.

In all cases of insecticide intoxication associated with typical EEG changes which could be followed up, EEG patterns returned to normal. In the case of pre-existent atypical changes the EEG's returned to their former patterns. These latter workers with pre-existent atypical EEG changes were not more susceptible to insecticide intoxication than workers with previously normal patterns. When they had insecticide intoxications these occurred at approximately the same blood levels as was the case in the group with normal pre-exposure EEG's.

From our experience and on theoretical grounds we can schematically classify organochlorine insecticide intoxications into 3 types:

Type 1: an acute convulsive intoxication with little or no prodomi following acute gross over-exposure;

Type 2: a cumulative intoxication due to regularly repeated smaller doses. Non-specific but consistently pertinent preconvulsive signs and symptoms may be a manifestation of this type of intoxication which may or may not culminate in a convulsion;

Type 3: an acute convulsive intoxication following an in itself insignificant overdose superimposed on an accumulative subclinical intoxication.

The mean insecticide levels in the blood at the moment of insecticide intoxication were 0.320 μg/ml, or 533 times the present D.Eq. of the general population in the U.K. and the U.S.A.

Chapter 17

Patterns of absenteeism due to disease and accidents

17.1   INTRODUCTION

Hoogendam *et al.* (1965) reported on the sickness absenteeism rates in our insecticides workers as compared with those in oil refinery and other chemical plant workers. The results of this study are tabulated again in Table 34.

From this table it is clear that sickness absenteeism in our insecticide workers is of the same order as sickness absenteeism in our other workers. Also, the pattern of sickness absenteeism of the insecticide workers did not differ significantly from that of their non-exposed colleagues.

Hayes *et al.* (1968) in their study of occupationally exposed workers wrote: "It is noteworthy that there was no meaningful correlation between the use of sick leave and the concentration of dieldrin and related compounds in samples measured in this study".

It is now six years since the data mentioned in the study of Hoogendam *et al.* (1965) were compiled, and thus it is appropriate to study once more the pattern and rate of absenteeism due to disease and the rate of absenteeism due to accidents. The accident rate is so low that a comparison of the accident pattern would not be meaningful. Since no insecticide intoxications now occur, the overall accident pattern (at work as well as at home) does not differ from that of other chemical or oil refinery plant operators.

Before we report the available data, it should be mentioned here that the data as such have no absolute value and may not always be comparable. In the first place the

Table 34

SICKNESS ABSENTEEISM RATES, INSECTICIDE   WORKERS AND OTHER OIL REFINERY/ CHEMICAL PLANT WORKERS

8-yr average of total personnel (6000 men)	3.1%
8-yr average of all insecticide workers	2.8%
Average of insecticide workers with continuous exposure over 9 yr	1.99%
Average of insecticide workers after medical transfer	2.8%
Average of insecticide workers after convulsive intoxication	2.85%

average age of a plant population has to be taken into account and this is not the same for all plants. Van der Heiden and De Jonge (1968) examined the effect of age on the rate of disease in the male working population of our plants and found that length of absenteeism due to disease shows a rapid increase in older age groups. This is illustrated in Table 35.

Table 35

DAYS OF ABSENCE PER MAN, PER ANNUM IN OUR MALE EMPLOYEE GROUP

Age group	1966 and 1967	
	CAO	NON-CAO*
15 - 19	6.563	–
20 - 24	7.798	6.660
25 - 29	9.009	5.126
30 - 34	9.515	6.224
35 - 39	11.729	5.335
40 - 44	14.754	8.468
45 - 49	16.041	9.825
50 - 54	28.556	16.453
55 - 59	42.144	30.136

*CAO means Collective Labour Contract

A second factor which has to be taken into account is that our workers, before they return to work after disease or accident, have a medical examination. On many occasions operators and maintenance workers are not yet considered fit for their own work, whereas they would have been regarded fit for administrative work. The work situation carries still more weight in this decision for those workers who handle toxic materials or who work in other hazardous situations.

This cautious attitude of our medical department will certainly have an effect on the rate of absenteeism of our insecticide workers as compared with all other operators, in the same way as we know it to have an effect on the absenteeism rate of operators as compared with administrative workers. This factor will mainly affect the length of each spell of disease or accident, rather than its frequency.

17.2 PATTERN OF ABSENTEEISM DUE TO DISEASE

In Tables 36-39 the rates of absenteeism for the years from 1966 to and including 1969, are given for the various disease code groups in days per man-year. These data are given separately for the operators of the insecticide plant (CIS) and the various other chemical plants (A-J), and also for the total of all operators in our chemical industry (SNC). Only company employed workers have been entered in these tables. In Table 40 the data of these 4 years have been combined.

Table 36

ABSENTEEISM DUE TO DISEASE PER MAN-YEAR – 1966

Disease Code	A	B	CIS	C	D	E	F	G	H	I	J	TOTAL SNC
001-138	0.061	0.157	0.566			1.328	0.563	0.229				0.288
140-239		1.861										0.229
240-289			0.090	0.220								0.030
290-299							0.048					
300-326	0.209	0.330	0.098	0.260		0.361	0.071	0.057	0.226	0.069		0.140
330-398			0.115	0.040					0.081	0.103		0.051
400-468	0.365	0.496	0.975	1.460			0.690	0.095		1.905		0.591
470-475	0.470	0.904	3.426	0.960	0.500	0.508	1.929	0.629	0.758	0.957		1.061
480-484	1.226	1.539		1.880	1.953	0.426	0.056	2.676	1.484	1.810		1.630
490-502	0.417	0.061		0.140		2.672	0.175	0.352		0.164		0.134
510-527	0.391	0.356	0.246		0.234							0.163
530-531		0.096	0.033				0.008	0.010	0.016			0.035
571	0.583	0.426	0.402	0.120	0.125	0.066	0.524	0.600	0.774	0.052		0.472
532-587 (e 571)	0.452	1.174	0.475	0.540	1.375	0.443	0.714	0.476	0.016	0.328		0.589
602	0.035	0.122				0.279	0.310			0.517		0.074
590-637 (e 602)	0.017	0.252	0.164	0.240	0.188		0.143	0.067	0.016			0.085
690	0.374	0.061	0.189	0.300	0.063		0.079					0.073
691-716	0.200	0.087	0.336	0.140	0.063		0.024	0.038				0.198
726	0.452	1.591	0.803	2.520	0.516	0.574	0.476	1.657	0.226	0.509		0.935
725-749 (e 726)			0.016	1.500		1.377	0.571		0.774	0.819		0.318
784			0.033		0.047					0.112		0.007
790-791	0.165	0.017		0.020	0.344		0.444	0.095		0.293		0.154
780-795 (e 784, 790, 791)	0.565	0.522	0.786	0.640	0.156	0.262	0.500	0.086	0.371	0.052		0.406
Total	5.983	10.052	8.754	10.980	5.563	8.295	7.325	7.067	4.742	7.690		7.665
No. of workers	115	115	122	50	64	61	126	105	62	116		936

Table 37

ABSENTEEISM DUE TO DISEASE IN DAYS PER MAN-YEAR – 1967

Disease Code	A	B	CIS	C	D	E	F	G	H	I	J	Total SNC
001-138		2.158			0.432		0.333	0.171	0.377			0.353
140-239		1.167			0.103							0.133
240-289				0.060								0.010
290-299			0.180						0.033	0.041		0.020
300-326	1.132	0.351	0.180	0.075	0.155	0.331	0.127		0.066			0.218
330-398	0.579	0.772	1.008	0.642	0.123	0.485	0.119		0.311	0.203		0.360
400-468	0.921	0.263	0.287	0.493		1.015	0.984	0.981	0.967	2.179		0.600
470-475	1.018	0.526	1.172	0.612	1.077	1.081	0.857	0.705	0.967	0.967		0.927
480-484	1.544	1.421	1.451	1.448	1.110	0.066	0.095	1.010	1.148	1.130		1.206
490-502	0.018		0.164	0.313		0.184	0.111	0.133		0.293		0.102
510-527		0.202	0.098			0.022	0.048	0.562		0.106		0.130
530-531	0.053	0.132	0.139	0.119	0.097	0.507	0.881	0.114	0.098	0.122		0.092
571	0.719	0.807	0.754	0.985	0.342	0.860	0.381	0.657	0.459	0.472		0.641
532-587 (e 571)	0.167	0.009	0.197	0.776	0.394				0.426	0.008		0.311
602		0.070			0.316		0.437	0.086				0.059
590-637 (e 602)	0.009	0.070	0.180		0.458	0.110	0.071	0.133				0.159
690			0.074		0.052					0.057		0.037
691-716	0.272		0.156	0.627		0.147	0.032	0.124				0.115
726	0.158	0.693	0.189	0.597	0.774	0.154	0.635	0.667	0.885	0.846		0.542
725-749 (e 726)		2.088	1.041	0.090		0.338	0.151		0.262			0.402
784	0.851		0.164	0.269	0.071	0.007	0.095		0.066	0.057		0.151
790-791	0.035	0.079	0.090	0.254	0.161	0.088		0.448	0.262	0.431		0.173
780-795 (e 784, 790, 791)	0.175	0.053	0.500	0.642	0.419	0.375	0.397	0.029	0.082	0.106		0.282
Total	7.649	10.860	8.025	8.000	6.084	5.772	5.754	5.819	5.443	7.016		7.020
No. of workers	14	114	122	67	155	136	126	105	61	123		1123

Table 38
ABSENTEEISM DUE TO DISEASE IN DAYS PER MAN-YEAR – 1968

Disease Code	A	B	CIS	C	D	E	F	G	H	I	J	Total SNC
001-138	0.008	0.170			0.787	1.789				1.206		0.437
140-239		0.205										0.021
240-289	0.441		0.032	0.181								0.062
290-299												
300-326		0.938	0.250	0.042	0.206	0.110	0.839	0.273	0.125	0.635		0.364
330-398		0.188	1.460	0.014		0.073	0.065	1.083		0.294		0.346
400-468	0.008	0.080		3.083			1.702					0.397
470-475	1.542	1.473	1.145	0.972	1.187	0.954	0.871	0.719	1.286	0.897		1.098
480-484	1.347	2.304	2.306	1.875	2.639	1.706	1.758	1.488	3.589	2.802		2.135
490-502	0.398	0.402	0.274	0.278		0.156	0.153	0.273	0.393			0.212
510-527		0.866	0.081		0.103	0.358	0.129					0.159
530-531	0.093	0.241	0.145	0.055	0.103	0.275	0.065	0.050		0.032		0.111
571	0.542	0.464	0.516	1.264	0.671	0.890	1.000	0.281	0.464	0.929		0.692
532-587 (e 571)	0.932	0.071	0.532	0.458	0.265	0.073	0.113	0.331	0.482	0.111		0.316
602		0.196			0.006		0.581					0.092
590-637 (e 602)			0.234	0.403	0.194				0.125	0.151		0.102
690			0.177			0.037		0.025		0.032		0.030
691-716	0.102		0.685	0.583			0.903		0.125	0.063		0.238
726	0.440	0.482	0.323	0.153	0.690	0.679	0.702	0.554	0.071	0.587		0.510
725-749 (e 726)	0.856	0.241	2.202	0.611	0.897	0.853						0.606
784	0.254		0.218	0.097	0.077		0.226	0.025	0.089	0.063		0.097
790-791	0.169	0.009	0.806		0.084	0.009	0.282	0.504		0.437		0.267
780-795 (e 784, 790, 791)	0.203	0.188	0.105	0.931	0.387	0.257	0.508	0.322	0.107	0.286		0.320
Total	7.339	8.518	11.492	11.000	8.297	8.220	9.895	5.926	6.857	8.524		8.623
No. of workers	118	112	124	72	155	109	124	121	56	126		1117

Table 39

ABSENTEEISM DUE TO DISEASE IN DAYS PER MAN-YEAR – 1969

Disease Code	A	B	CIS	C	D	E	F	G	H	I	J	Total SNC
001-138		0.200	0.115			0.320	0.205					0.077
140-239	1.034		0.298			0.320						0.165
240-289		0.036	0.077	0.039			0.213	0.040				0.077
290-299				0.105	0.014			0.160		0.254		0.020
300-326	0.138	0.809	0.202	0.066	0.297	0.013	0.066					0.109
330-398	0.293	0.355	0.442	1.303	0.878		0.311	0.648				0.370
400-468		0.236	1.327	2.211	1.027		1.000					0.591
470-475	1.043	1.718	4.279	1.434	2.419	1.387	1.230	1.120	2.481	0.341	0.577	1.163
480-484	2.190	3.364	0.260	4.447	0.311	2.667	2.852	4.480	0.865	0.135	0.385	2.956
490-502	0.534	0.064		0.408			0.279	0.048	2.962	0.738		0.211
510-527			0.500				0.148	0.240	0.288	1.230		0.093
530-531	0.112	0.100	0.462		0.020	0.013		0.120	0.077	0.040	0.154	0.096
571	0.405	0.400	0.654	0.355	0.507	0.307	0.566	0.752	0.231	0.794	1.154	0.545
532-587 (e 571)	0.345		1.038		0.257	1.013	0.393			0.413		0.335
602				0.118	0.135		0.525					0.068
590-637 (e 602)				0.329	0.115			0.848				0.140
690		0.200	0.067		0.155		0.016		0.096			0.063
691-716	0.586	0.109	0.096	0.382	0.399		0.189	0.056	0.173	0.063	0.462	0.161
726	0.198	0.145	0.269	0.526		1.160	0.434	1.120	0.750			0.523
725-749 (e 726)	0.121	0.109	0.337		0.081	0.133	0.246	0.280	1.077	0.468	0.808	0.178
784	0.224	0.236	0.067				0.016		0.135			0.086
790-791	0.216	0.555	0.808	0.434	0.230	0.280	0.180	0.176	0.712	0.103		0.317
780-785 (e 784, 790, 791)	0.103	0.100	0.288	0.066	0.284	0.413	0.779	0.160	0.058	0.151	0.115	0.250
Total	7.543	8.736	11.587	12.224	7.128	8.027	9.648	10.248	9.904	4.730	3.731	8.605
No. of workers	116	110	104	76	148	75	122	125	52	126	26	1080

Table 40

ABSENTEEISM DUE TO DISEASE IN DAYS PER MAN-YEAR – 1966-1969

Disease Code	A	B	CIS	C	D	E	F	G	H	I	J	Total SNC
001-138	0.015	0.676	0.172		0.362	0.787	0.277	0.092	0.100	0.310		0.291
140-239	0.261	0.820	0.066	0.011	0.031	0.063						0.133
240-289	0.112	0.009	0.049	0.136			0.064	0.011	0.009	0.075		0.045
290-299			0.047		0.004		0.044					0.010
300-326	0.365	0.603	0.138	0.098	0.107	0.089	0.225	0.072	0.091	0.230		0.212
330-398	0.216	0.328	0.718	0.547	0.121	0.142	0.143	0.478	0.039	0.187		0.291
400-468	0.320	0.271	0.424	1.872	0.249	0.255	0.699	0.248	0.641	1.031		0.543
470-475	1.022	1.149	1.782	1.011	1.025	0.976	0.942	0.805	0.965	0.888	0.577	1.061
480-484	1.577	2.144	1.924	2.506	2.038	1.827	1.841	2.471	2.238	1.745	0.385	1.987
490-502	0.343	0.131	0.172	0.298	0.088	0.068	0.145	0.197	0.160	0.112		0.165
510-527	0.097	0.357	0.220		0.059	0.168	0.141	0.195		0.026		0.136
530-531	0.065	0.142	0.184	0.068	0.065	0.099	0.030	0.075	0.048	0.061	0.154	0.086
571	0.562	0.525	0.578	0.796	0.460	0.567	0.743	0.570	0.494	0.637	1.154	0.593
532-587 (e 571)	0.477	0.319	0.542	0.321	0.437	0.551	0.402	0.197	0.234	0.259		0.379
602	0.009	0.098		0.079	0.096	0.021	0.351	0.019				0.073
590-637 (e 602)	0.002	0.064	0.150	0.204	0.255	0.039	0.147	0.263	0.035	0.039		0.123
690	0.004	0.082	0.129	0.057	0.056	0.010	0.042	0.037	0.022	0.039	0.462	0.052
691-716	0.333	0.049	0.328	0.453	0.052	0.052	0.285	0.037	0.130	0.136		0.177
726	0.251	0.736	0.400	0.819	0.611	0.570	0.562	0.989	0.628	0.676	0.808	0.615
725-749 (e 726)	0.361	0.614	0.926	0.472	0.266	0.612	0.243	0.077	0.312	0.026		0.380
784	0.330	0.058	0.123	0.068	0.073	0.003	0.084	0.007	0.048	0.057		0.089
790-791	0.147	0.162	0.413	0.219	0.180	0.089	0.227	0.307	0.251	0.289	0.115	0.230
780-795 (e 784, 790,791)	0.261	0.217	0.424	0.555	0.339	0.331	0.544	0.156	0.160	0.151	0.077	0.311
Total	7.130	9.554	9.909	10.589	6.973	7.320	8.137	7.349	6.602	6.976	3.732	7.982
No. of workers	116	113	118	66	131	95	125	114	58	123	26	1085

A description of the disease code groups — according to the "International Statistical Classification of Diseases, Injuries and Causes of Death, 1948" — is given in Table 41.

A statistical analysis of the data in Tables 36-40 (De Jonge, 1970) showed that in plants B, C and in the insecticide plant the total rate of absenteeism was significantly higher than in the other plants. The reason for this difference could not be given from these data.

An analysis of variance showed that there was no significant interaction between plants and disease code groups, or in other words, no disease code group occurred more or less frequently in any one plant.

The only comment upon this statistic is that the rather high rates in the insecticide plant in 1967 and 1968 for diagnose group 330-398 (nervous system and sense organs)

Table 41
MEANING OF DISEASE CODE

Disease code	Meaning
001-138	Infective and parasitic diseases
140-239	Neoplasms
240-289	Allergic, endocrine system. metabolic and nutritional diseases
290-299	Diseases of blood and blood-forming organs
300-326	Mental, psychoneurotic, personality
330-398	Nervous system and sense organs
400-468	Circulatory system
470-475	Acute upper respiratory infections
480-484	Influenza
490-502	Pneumonia and bronchitis
510-527	Other diseases of respiratory system
530-531	Dental diseases
571	Gastroenteritis and colitis
532-587 (e 571)	Other diseases of digestive system
602	Renal calculus
590-637 (e 602)	Other diseases of genitourinary system
690	Boils and carbuncles
691-716	Other diseases of skin and cellular tissue
726	Low back pain
725-749 (e 726)	Other diseases of bones and organs of movement
784	Symptoms referable to upper gastrointestinal system
790-791	Nervousness and debility
780-795 (e 784, 790, 791)	Other ill defined conditions

were due to two cases of mastoidotomy following chronic otitis media with an absenteeism of 198 and 54 days respectively.

17.3 RATE OF ABSENTEEISM DUE TO DISEASE

The different relevant data for absenteeism due to disease have been tabulated in Table 42, and the following rates were calculated and included in this table:
a. disability rate: the average duration of absenteeism per man per year;
b. frequency rate: the total spells of absenteeism per 100 man per year;
c. severity rate: the average duration of each spell of absenteeism.

The interrelation between these rates is shown in Tables 42 and 43. In Table 42 these data of the insecticide operators are compared with corresponding data of all operators working in our chemical plants.

In 17.2 it had already been shown that the rate of absenteeism in (amongst others) the insecticide plant operators was significantly higher than it was in most other chemical plants. From Table 42 it can be calculated that in insecticide workers as compared with all chemical operators:
— the disability rate is 1.24 times higher,
— the frequency rate is 1.08 times higher, and
— the severity rate is 1.14 times higher.

17.4 RATE OF ABSENTEEISM DUE TO ACCIDENT

Data and rates of absenteeism due to accident have been calculated and tabulated in Table 43 in the same way as this was done in 17.3 for absenteeism due to disease. All accidents were taken into account: at work, at home, as well as traffic accidents.

Insecticide workers compared with all chemical plant operators have lower disability and severity rates than the overall group and a frequency rate which is not significantly higher.

17.5 DISCUSSION

A significantly higher absenteeism due to disease was found in insecticide workers compared with all chemical plant operators. On the other hand, no disease code group occurred significantly more or less frequently in any of the plants, and the higher absenteeism was to a large extent due to an increased length of each spell of disease as expressed in the severity rate. This in its turn was due to extra caution on the part of our doctors rather than to infirmity of our insecticides workers.

In 17.1 the increase of the disability rate with increasing age was mentioned. The average age of our insecticide operators (not to be confused with the average age of all insecticide workers) was 35.4 years in 1969, whereas the average age of all operators working in our chemical plants was 31.5 years in the same year. This difference of nearly 4 years in average age of workers in the CAO group would account, according to the exponential curve representing the relation given in Table 35, for a $\frac{11.0}{9.6} = 1.15$ times higher disability rate in our insecticide operators as compared with all chemical operators, whereas actually we found a factor 1.24.

Table 42
DISEASE RATES OF INSECTICIDE OPERATORS COMPARED TO ALL CHEMICAL PLANT OPERATORS

Year	Insecticide plant operators						All chemical plant operators					
	No. empl.	Total days absent	Disability rate	Total spells	Frequency rate	Severity rate	No. empl.	Total days absent	Disability rate	Total spells	Frequency rate	Severity rate
	(A)	(B)	(B/A)	(C)	$\left(\frac{C\times100}{A}\right)$	(B/C)	(A)	(B)	(B/A)	(C)	$\left(\frac{C\times100}{A}\right)$	(B/C)
1966	122	1068	8.8	130	106.6	8.2	936	7174	7.7	957	102.4	7.5
1967	122	979	8.0	120	98.4	8.2	1123	7884	7.0	992	88.3	7.9
1968	124	1425	11.5	129	104.0	11.0	1117	9621	8.6	1193	106.8	8.1
1969	104	1205	11.6	144	138.5	8.4	1080	9293	8.6	1208	111.1	7.7
Total 4 years	-	4677	-	523	-	-	-	33972	-	4350	-	-
Average 4 years	118	1169	9.9	131	111.0	8.9	1064	8493	8.0	1088	102.3	7.8

Table 43

ACCIDENT RATES OF INSECTICIDE OPERATORS COMPARED TO ALL CHEMICAL PLANT OPERATORS

Year	Insecticide plant operators						All chemical plants operators					
	No.empl.	Total days absent	Disability rate	Total spells	Frequency rate	Severity rate	No.empl.	Total days absent	Disability rate	Total spells	Frequency rate	Severity rate
	(A)	(B)	(B/A)	(C)	$\left(\dfrac{C \times 100}{A}\right)$	(B/C)	(A)	(B)	(B/A)	(C)	$\left(\dfrac{C \times 100}{A}\right)$	(B/C)
1966	122	137	1.12	18	14.8	7.6	936	1372	1.47	116	12.4	11.9
1967	122	24	0.20	6	4.9	4.0	1123	1361	1.21	118	10.5	11.5
1968	124	242	1.95	18	14.5	13.4	1117	1618	1.45	111	9.9	14.6
1969	104	68	0.65	9	8.7	7.6	1080	972	0.90	90	8.4	10.8
Total 4 years	-	471	-	51	-	-	-	5323	-	435	-	-
Average 4 years	118	117.75	1.00	12.75	10.8	9.3	1064	1330.75	1.25	108.75	10.2	12.2

Then the second factor mentioned in 17.1, which cannot be quantified, has not yet been taken into account. This really means that absenteeism due to disease in our insecticide operators is equal, if not lower, than that of a comparable group of operators.

17.6   SUMMARY

Absenteeism due to disease and accident of our insecticide workers was compared with the same data of all chemical plant operators for the years 1966, 1967, 1968, 1969 and for these four years taken together.

Taking into account the difference in average age of both groups, there is no significant difference in disability, frequency and severity rate between the two groups, both for diseases and for accidents.

Moreover there was no disease group which occurred significantly more or less frequently in any of our chemical plants including insecticides compared with each other.

Chapter 18

Parameters as related to the insecticide level in the blood

18.1   INTRODUCTION

In chapters 14 and 15 certain parameters occurring in long-term and extreme exposure groups have been grouped and discussed. In this chapter these parameters are correlated with the blood level of dieldrin equivalent (D.Eq.) measured in the same blood sample.

Eight parameters will be discussed. The values of the various parameters are plotted on the vertical, against the D.Eq. — divided in nine level groups — on the horizontal line; every parameter in each blood level group according to the scheme next to this text:

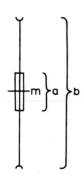

m = the geometric mean of all observations in this category

a = the 95% confidence range of this geometric mean

b = the 95% confidence range of all observations in this category.

All available data from all insecticide workers — irrespective length of exposure — were taken, up to a maximum of 100 per parameter level group. When more data were available 100 were chosen at random. The nine categories of D.Eq. levels were chosen in such a way that the threshold below which no signs or symptoms of intoxication were observed *i.e.* D.Eq. 0.20, is in the middle of the range. Of course most available data are below this threshold level. For some parameters, however, enough data above this threshold level are on record to give an indication of the trend at higher exposure levels.

All blood samples have been taken in a non-fasting state, usually between 8 and 11 a.m.

18.2   PARAMETERS

Serum alkaline phosphatase

The results of the serum alkaline phosphatase level determinations are given in Table 44 and Fig. 15. The normal 95% range for the method used is 1.06-2.68 mMolU, with an average of 1.87 mMolU.

All results below the 0.20 D.Eq. threshold level are normal, although the top of the 95% confidence range is slightly higher than the norm given in nearly all groups. An analysis of variance demonstrated a significant increase ($p < 0.001$) of the mean values in the 0.21-0.30 D.Eq. groups. Also, the 95% confidence range in these groups is wider. At 0.31 D.Eq. and higher there is again a tendency to a lower mean and scatter, although the number of observations here is too small to draw conclusions.

Thus, although no pathological or abnormal high levels were found, there is a

Table 44

ALKALINE PHOSPHATASE, SGOT, SGPT AND LDH AT DIFFERENT EXPOSURE LEVELS

Parameter	D.Eq. group	No.	Range	95% conf. range	Geom. mean	95% conf. range of geom. mean
				All observations		
Alkaline	⟨0.01	100	0.9-4.8	1.06-3.04	1.79	1.78-1.90
phosphatase	0.01-0.05	100	0.9-3.0	0.98-3.01	1.70	1.60-1.81
in mMol∪	0.06-0.10	100	0.9-3.3	0.97-2.65	1.62	1.55-1.72
	0.11-0.15	100	0.9-3.0	1.04-2.80	1.70	1.64-1.79
	0.16-0.20	100	1.0-3.6	1.06-3.14	1.81	1.70-1.91
	0.21-0.25	47	1.1-3.1	1.17-3.58	2.05	1.89-2.23
	0.26-0.30	13	1.3-2.8	1.46-3.36	2.19	1.98-2.45
	0.31-0.40	14	1.3-2.5	1.26-2.91	1.93	1.74-2.13
	⟩0.40	5	1.4-2.1	1.18-2.93	1.86	1.52-2.28
SGOT in K.U.	⟨0.01	100	8-34	8.7-24.3	14.4	13.5-15.3
	0.01-0.05	100	6-26	6.8-23.5	12.6	11.6-13.6
	0.06-0.10	100	5-36	7.1-30.2	14.9	13.8-16.0
	0.11-0.15	100	5-44	7.0-30.3	14.6	13.5-15.8
	0.16-0.20	29	6-32	7.4-35.4	16.4	14.2-18.8
	0.21-0.25	15	9-31	9.5-36.3	18.6	15.7-21.9
	0.26-0.30	11	8-32	6.2-38.0	15.3	11.6-20.4
	0.31-0.40	12	10-23	10.3-28.3	17.0	14.5-19.7
	⟩0.40	5	12-21	9.5-35.0	18.2	13.6-24.4
SGPT in K.U.	⟨0.01	100	4-35	5.1-30.8	12.3	11.2-13.4
	0.01-0.05	100	5-48	4.7-36.2	13.1	11.6-14.6
	0.06-0.10	100	1-40	4.4-44.3	14.1	12.5-15.7
	0.11-0.15	64	2-40	4.5-36.3	12.9	11.4-14.5
	0.16-0.20	26	5-34	4.4-50.5	15.0	12.1-19.2
	0.21-0.25	13	5-33	3.7-54.4	14.2	9.8-20.6
	0.26-0.30	10	5-23	3.6-36.4	10.4	6.7-15.4
	0.31-0.40	11	7-31	5.5-44.8	15.9	11.6-21.8
	⟩0.40	3	15-20	9.3-32.0	17.2	12.0-24.6
LDH in mU/ml	⟨0.01	42	75-317	69-297	146	129-162
	0.01-0.05	71	83-350	75-295	151	138-162
	0.06-0.10	72	83-333	78-293	150	139-163
	0.11-0.15	34	83-250	79-293	152	134-171
	0.16-0.20	10	100-233	68-245	129	107-158
	0.21-0.25	0	–	–	–	
	0.26-0.30	0	–	–	–	
	0.31-0.40	0	–	–	–	
	⟩0.40	0	–	–	–	

significantly higher, but still quite normal, mean of the serum alkaline phosphatase levels in the exposure groups between 0.21 and 0.30 D.Eq. The top of the range is slightly higher than the norm in all groups.

SGOT

The SGOT levels are also given in Table 44 and Fig. 15. The normal range for the method used is 8-40 K.U. (Karmen units). Means and ranges are normal in all groups.

176

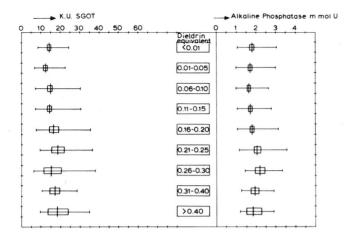

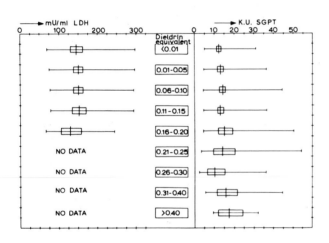

Fig. 15. Alkaline phosphatase, SGOT, SGPT and LDH at different exposure levels.

With an analysis of variance the 0.01-0.05 D.Eq. exposure group appeared to be significantly lower than the other groups ($p < 0.001$). With the Spearman rank correlation test a significant correlation was found in the sense that SGOT increases with D.Eq. exposure.

S G.P T

The normal range for the method used is 5-35 K.U. Although a few turbid sera, due to a non-fasting state, gave higher results, the means in all exposure groups do not vary significantly, as is seen from its 95% confidence range in Table 44 and Fig. 15. This was confirmed by an analysis of variance. With the Spearman rank correlation test a slight, but significant, increase of SGPT with increasing D.Eq. was found.

Table 45

TOTAL SERUM PROTEIN, HAEMOGLOBIN, WHITE BLOOD CELLS AND SEDIMENTATION
RATE OF ERYTHROCYTES AT DIFFERENT EXPOSURE LEVELS

Parameter	D. Eq. group	All observations				95% conf. range of geom.mean
		No.	Range	95% conf. range	Geom mean	
Total serum protein in g/l	⟨0.01	42	60.5-82.5	61.1-82.0	71.1	69.2-72.7
	0.01-0.05	71	61.5-87.0	61.2-80.4	70.7	69.3-71.7
	0.06-0.10	72	58.0-78.0	61.1-78.2	69.1	67.9-70.6
	0.11-0.15	33	59.5-81.5	60.4-78.6	69.2	67.4-70.6
	0.16-0.20	10	62.0-85.5	57.1-80.3	71.7	66.8-77.0
	0.21-0.25	0	–	–	–	–
	0.26-0.30	0	–	–	–	–
	0.31-0.40	0	–	–	–	–
	⟩0.40	0	–	–	–	–
Haemoglobin in %	⟨0.01	100	78-106	80.8-106.2	92.7	91.3-94.1
	0.01-0.05	100	82-105	81.1-104.7	92.0	90.7-93.3
	0.06-0.10	100	81-105	82.8-104.7	93.2	92.0-94.3
	0.11-0.15	100	80-108	81.2-106.6	93.2	91.7-94.3
	0.16-0.20	100	73-105	83.0-106.2	94.0	92.8-95.2
	0.21-0.25	52	83-108	86.2-105.9	95.3	93.7-96.7
	0.26-0.30	26	84-104	85.4-106.6	95.2	93.3-97.4
	0.31-0.40	21	85-101	83.5-106.2	94.2	91.7-96.6
	⟩0.40	12	83-106	80.9-112.1	95.3	90.9-99.8
WBC's mm³	⟨0.01	100	3.500-12.100	3.715-10.186	6.060	5.789-6.412
	0.01-0.05	100	3.000-11.000	3.320-10.099	5.972	5.644-6.337
	0.06-0.10	100	3.800-9.600	3.611-9.185	5.691	5.408-6.008
	0.11-0.15	100	3.700-10.700	3.724-10.393	6.222	5.875-6.472
	0.16-0.20	100	3.600-9.500	3.768-9.010	5.743	5.480-6.092
	0.21-0.25	52	3.500-9.400	3.210-10.192	5.690	5.295-6.211
	0.26-0.30	26	3.500-10.300	3.218-11.005	5.842	5.218-6.649
	0.31-0.40	21	3.900-10.300	3.255-11.280	5.991	5.188-6.667
	⟩0.40	12	4.500-9.400	3.787-9.376	5.959	5.228-6.972
SRE in mm 1st hour	⟨0.01	100	1-11	0.7-10.8	2.7	2.2-3.3
	0.01-0.05	100	1-14	0.7-12.4	2.7	2.2-3.3
	0.06-0.10	100	1-55	0.6-16.7	3.2	2.7-3.8
	0.11-0.15	100	1-28	0.8-19.0	3.9	3.2-4.8
	0.16-0.20	100	1-23	0.8-15.6	3.7	3.0-4.3
	0.21-0.25	52	1-40	0.5-19.7	3.2	2.4-4.2
	0.26-0.30	26	1-7	0.6-7.8	2.2	1.5-2.8
	0.31-0.40	21	1-8	1.0-9.3	3.1	2.4-4.0
	⟩0.40	12	1-9	0.5-15.4	2.7	1.7-4.5

LDH

Serum LDH activity is considered to be normal up to 300 mU/ml. For LDH only
data for exposures up to 0.20 D.Eq. are available, since, when we started to measure LDH
levels, D.Eq. levels exceeding 0.20 D.Eq. were no longer encountered. Results of LDH
determinations are shown in Table 44 and Fig. 15 again. All results in all groups are
normal. In the groups available an analysis of variance did not demonstrate significant

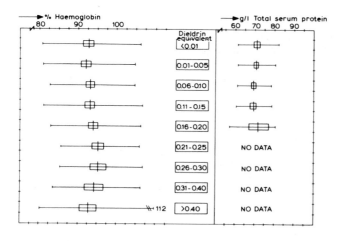

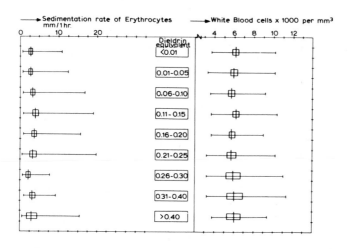

Fig. 16. Total serum protein, haemoglobin, white bloodcells and sedimentation rate of erythrocytes at different exposure levels.

differences between groups. With the Spearman rank correlation test no correlation between mean levels and D.Eq. exposure was found.

Total serum proteins

Here again, and for the same reasons as with LDH serum levels, only data of exposures below the threshold level of 0.20 D.Eq. are available. See Table 45 and Fig. 16.

The results are normal, and an analysis of variance does not show significant differences between group means. With the Spearman rank correlation test no correlation was found between total serum protein and D.Eq. level.

Haemoglobin

Table 45 and Fig. 16 also give the data for haemoglobin. Ranges and means are normal for all groups. An analysis of variance demonstrates a significant difference between group means (p = 0.01). With the Spearman rank correlation test the increase of haemoglobin with increasing D.Eq. is significant.

White blood cells

Again in Table 45 and Fig. 16 the results of white blood cell counts are given. The group means do not differ significantly as is shown by an analysis of variance. No correlation of white blood cells with D.Eq. level was found by the Spearman rank correlation test.

Sedimentation rate of erythrocytes

The high SRE's occasionally encountered are found in those workers with pre-existent diseases such as rheumatoid arthritis, as discussed in chapter 14.1 and 14.2.3, and clearly are scattered over the various D.Eq. groups and recorded at different times. Nearly all SRE's, however, fall in the normal range. Group means are quite normal and not related to the D.Eq. exposure group, as was shown with the Spearman rank correlation test. An analysis of variance did not demonstrate significant differences between groups. Results of SRE's are given in Table 45 and Fig. 16.

18.3   DISCUSSION OF RESULTS

For LDH serum levels and total serum proteins only data below the intoxication threshold level of 0.20 μg/ml D.Eq. were available. In all other parameters: serum alkaline phosphatase, SGOT, SGPT, haemoglobin, WBC's and SRE many data from exposures higher than this threshold level could be compiled as well.

In serum alkaline phosphatase a statistically significant increase was found, occurring in the 0.21-0.30 D.Eq. exposure groups, where our intoxication threshold was exceeded. Although means remained quite within the normal range, a few samples exceeding the normal range were found in practically all groups, mostly, however, in the lower exposure groups. Therefore, this is not likely to be related to the degree of exposure as expressed in the D.Eq. This is in agreement with the fact that in individual cases of over-exposure to these insecticides we found no increased serum alkaline phosphatase.

SGOT and SGPT showed normal means and ranges in all groups. No statistically significant differences between groups were demonstrated although in both cases, but more in SGOT than SGPT, there was a significant tendency to increase with higher D.Eq. levels. We find no explanation for the lower SGOT in the 0.01-0.05 D.Eq. group. The De Ritis quotient was not determined individually, and it would not be correct to calculate it on the group means, although this too would give perfectly normal results here.

In serum LDH and total serum proteins no differences between groups were found, and all results were normal.

Haemoglobin showed a statistically significant increase within normal ranges, with increasing D.Eq. level. The explanation for this phenomenon is not clear. In our opinion the absolute insecticide content of the blood is too small to ascribe this to an increased insecticide-carrier function of the haemoglobin. It might, however, be that the results in the higher D.Eq. groups, which are of course all results of earlier exposure years, were influenced more than the results of other groups by variations in the standard of haemoglobin determination. Whatever the explanation, we feel that a negative influence of these insecticides on the haemoglobin content of the blood may be excluded.

White blood cell counts and sedimentation rates or erythrocytes are not influenced by increased exposures to these insecticides as expressed in the D.Eq.

The finding of increased group means of serum alkaline phosphatase with a simultaneously increasing trend of SGOT and SGPT group means, although within normal limits, is strongly suggestive, as was discussed in chapter 2.5.4, of being a sign of the adaptation phase of the liver. Due to an increase in the size of the liver cells and the resultant slight intrahepatic cholestasis, serum alkaline phosphatase levels tend to increase during this phase. However, we found no enlarged livers at palpation in our high exposure groups.

We must conclude, therefore, that this effect — measured only in cases where our intoxication threshold of 0.20 D.Eq. was exceeded — is the first manifest adaptive response to the measured absorption, as far as examined in the parameters mentioned above.

18.4   SUMMARY

In this chapter eight parameters have been statistically evaluated and grouped according to the insecticide level measured at the same time.

Of the LDH and total serum proteins only data of cases below our intoxication threshold level of 0.20 D.Eq. were available. Means and 95% confidence ranges were normal, and there were no differences between groups.

WBC's and SRE were normal and not influenced by the D.Eq. level, and also in groups up to at least twice our threshold level.

Very slight increases within normal limits were found in SGOT, SGPT and haemoglobin with increasing D.Eq. levels.

Serum alkaline phosphatase levels, however, showed a statistically significant increase of its mean level in the 0.21-0.25 and the 0.26-0.30 μg/ml D.Eq. exposure groups, although means remained within normal limits. As was discussed, this response must be regarded as a first sign of adaptation to the measured absorption.

Chapter 19

Study of general health parameters of all insecticide workers

19.1 GROUPS OF WORKERS EXAMINED

In this study parameters of insecticide workers are compared with those of a group of workers who have had no occupational exposure to insecticides. Members of both groups are shift workers. Daytime workers are not considered here.

An appropriate control group was formed of some operators from a catalytic cracking unit and some from one of our plastic manufacturing plants, both of whom had routine medical examinations during the same period in which the examination of our insecticide workers was carried out, that is, the last three months of 1969.

Some of the work in the insecticide plant is still a batch process and this includes the formulation unit. For this reason more unskilled and contractor labour is employed in this plant and several of our skilled operators started in this category, as was discussed in chapters 10 and 12.2.

It was not possible to choose a control group of shift workers from the same demographic category: we had to use a control group of trained operators, without unskilled contractors.

The group of former insecticide workers and the subgroup of those long-term insecticide workers who have worked for more than 12 years in insecticides and are still working there, have about the same demographic composition as the group of insecticide workers. The difference in demographic background with the control group may also account for the higher top of the age range and the higher mean ages of these groups (groups IV, V, VI in Table 46 and Fig. 17).

All this resulted in 6 groups of workers:

Group I	Operators:	cracking and plastic manufacture; n = 133.
Group II	Operators:	aldrin and dieldrin manufacture; n = 40.
Group III	Operators:	endrin manufacture; n = 27.
Group IV	All workers:	insecticide plant; n = 113.
Group V	All workers:	> 12 year in insecticides; n = 32.
Group VI	All workers:	former insecticide workers; n = 83.

Age, total period in chemical industry in months, as well as the period of insecticide work in months of these groups are compared in Table 46 and Fig. 17; the time (in months) since the last insecticide exposure for group VI, is also given.

In Table 47 and Fig. 18 the insecticide levels in the blood are given: dieldrin and Telodrin separately, and combined, expressed as Dieldrin Equivalent (D.Eq.) (see chapter 2.3.3). Insecticide levels in Fig. 18 have been plotted on a logarithmic scale.

182

Table 46

AGE AND DURATION OF OCCUPATIONAL INSECTICIDE EXPOSURE

Parameter	Group	All observations		
		No.	Range	Geom. mean
Age in years	I	133	19 - 54	33.56
	II	40	19 - 53	34.75
	III	27	23 - 51	35.46
	IV	113	19 - 64	37.95
	V	32	35 - 61	44.30
	VI	83	24 - 65	37.07
Total operation period	I	133	14 - 264	97.56
in months	II	40	3 - 277	86.17
	III	27	39 - 261	105.00
	IV	113	3 - 279	92.30
	V	32	144 - 279	174.04
	VI	83	40 - 264	126.77
Total period of insecticide	I	133	0	0
exposure in months	II	40	3 - 177	70.62
	III	27	17 - 164	77.85
	IV	113	2 - 182	72.92
	V	32	144 - 182	159.70
	VI	83	3 - 138	39.89
Time since last insecticide exposure in months	VI	83	3 - 138	49.85

In the aldrin-dieldrin group (II), range and means of the dieldrin levels in the blood are lower than those than for 1969 of the aldrin-dieldrin workers given in chapter 13, Table 20. This is because many of the highest dieldrin blood levels are found in the daytime contractor-workers, who were not included in this group of shift workers, whereas intermediate workers with lower levels were considered in this study.

Endrin was not found in any of the blood samples examined at a detection threshold level of 0.005 $\mu g/ml$.

19.2   PARAMETERS USED IN THIS STUDY

Twenty hematological and biochemical parameters were determined.

Haemoglobin was determined by the cyano-haemoglobin method. White and red blood cells were counted by a Coulter counter, which also calculated MCV and hematocrit.

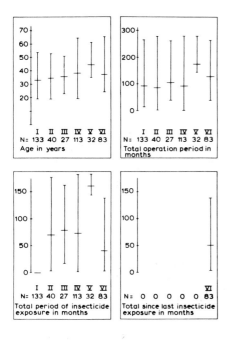

Fig. 17. Geometric mean and range of age, total operation period, period of insecticide exposure and total since last insecticide exposure.

For the sedimentation rate of erythrocytes first hour readings of the Westergren method were used.

Plasma cholinesterase was determined by the electrometric method described by Michel (1949).

SGPT was determined by the colorimetric method of Reitman and Frankel (1957).

All 12 other parameters were determined by our Technicon SMA 12/60 Auto-Analyzer, survey model. In this case the choice of the parameters and methods used was determined by those available on the Auto-Analyzer: calcium^{++}, inorganic phosphate, glucose, blood urea nitrogen (BUN), uric acid, cholesterol, total serum protein, albumin, total bilirubin, alkaline phosphatase, LDH and SGOT, all parameters determined in blood serum and registrated on a serum chemistry graph.

All examinations were carried out as part of a routine periodical medical examination in our industrial preventive medical programme. Workers could therefore not be in a fasting state and the time of the day when blood sampling was done varied between 8 a.m. and 4 p.m. Thus sampling was not done under quite standarized conditions. This may have influenced some individual data, but does not disturb the comparabilities of grouped data.

Table 47
INSECTICIDE LEVELS IN THE BLOOD

Parameter	Group	No.	Range	95% range	Geom. mean	95% conf. range of geom.mean
Dieldrin level in	I	133	⟨0.005	–	–	–
blood in μg/ml	II	40	⟨0.005-0.076	0.0009-0.1048	0.0095	0.0065-0.0139
	III	27	⟨0.005-0.015	0.0011-0.0073	0.0028	0.0024-0.0034
	IV	113	⟨0.005-0.149	0.0007-0.1124	0.0085	0.0067-0.0109
	V	32	⟨0.005-0.060	0.0010-0.0653	0.0081	0.0056-0.0118
	VI	73	⟨0.005-0.025	0.0009-0.0113	0.0032	0.0028-0.0038
Endrin level in blood in μg/ml	All groups: all workers ⟨0.005					
Telodrin level	I	133	⟨0.002	–	–	–
in blood in μg/ml	II	40	⟨0.002-0.008	0.00043-0.00963	0.00204	0.00160-0.00261
	III	27	⟨0.002-0.007	0.00043-0.00384	0.00129	0.00105-0.00159
	IV	113	⟨0.002-0.008	0.00043-0.00694	0.00172	0.00151-0.00196
	V	32	⟨0.002-0.008	0.00056-0.01180	0.00251	0.00193-0.00328
	VI	75	⟨0.002-0.031	0.00046-0.01641	0.00274	0.00223-0.00336
D.Eq. in blood in	I	133	⟨0.005	–	–	–
μg/ml	II	40	⟨0.005-0.105	0.0021-0.3060	0.0253	0.0170-0.0375
	III	27	⟨0.005-0.053	0.0005-0.0490	0.0042	0.0030-0.0074
	IV	113	⟨0.005-0.150	0.0010-0.3011	0.0173	0.0133-0.0226
	V	32	⟨0.005-0.120	0.0014-0.3729	0.0230	0.0141-0.0377
	VI	81	⟨0.005-0.310	0.0007-0.3360	0.0155	0.0110-0.0218

19.3   RESULTS

The results of all parameters examined were tabulated according to groups I-VI. Of each parameter group the number and the range of all observations were recorded, as well as the geometric means and the 95% confidence limits.

The results are tabulated in Tables 48-52 and illustrated in the corresponding Figs. 19-23, each table and figure containing the results of all groups of 4 parameters. In the figures the geometric means and the 95% confidence ranges of these means and of all observations are plotted as was described in chapter 18.1.

As hematocrit and MCV determination were not available from the start of this study, they could be determined for only part of the samples. In some other parameters a few results are missing as a result of technical mishaps.

19.4   DISCUSSION OF RESULTS

As regards haemoglobin and erythrocytes the means and ranges in all groups are within normal limits. The significantly lower values in some of the insecticide groups as compared to the control group are, in our opinion, an expression of the above mentioned

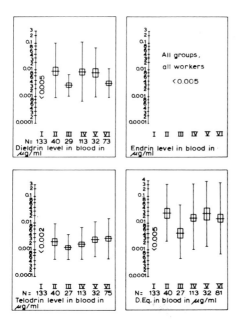

Fig. 18. Geometric mean and 95% confidence ranges of this mean and of all observations of the insecticide levels in the blood.

differences in demographic background between the groups. This is the more likely, as we found, *e.g.* an increase of Hb with increasing D.Eq. in chapter 18.

As regards hematocrit and MCV there are no significant differences between groups. The white blood cell count and the first hour reading of the sedimentation rate of erythrocytes are the same in all six groups considered, as are plasma cholinesterase and SGPT. All results are within normal ranges.

Normal calcium^{++} range for the method used is 8.5-10.5 mg%, for inorganic phosphate it is 2.5-4.5 mg% P. For all groups both parameters give normal values, although Ca^{++} is significantly higher and inorganic phosphate significantly lower in the control group. This may be explained by the relationship of calcium^{++} to total protein and albumin content of the blood serum. One gram of protein (mainly albumin) binds about 0.8 mg of calcium. When total protein and albumin are decreased, so is the calcium (Reece 1968).

Normal ranges for total serum protein and albumin for the method used are 6-8 g% and 3.5-5.0 g% respectively. All means and the distribution in all groups are normal and are actually in the upper ranges. Possibly due to the difference in demographic background serum proteins are higher in the control group, than in the other groups. This may account for the higher Ca^{++} and consequently lower inorganic phosphate found in the control group.

Normal range for glucose for the method used is 65-110 mg%, however as sampling was not done in a fasting state higher levels were to be expected. The means are normal

Table 48

HAEMOGLOBIN, RED BLOOD CELLS, HEMATOCRIT and MCV

| Parameter | Group | No. | All observations | | Geom. mean | 95 % conf. range of geom. mean |
			Range	95% conf. range		
Haemoglobin in %	I	133	74-110	82.5-104.1	92.69	91.77-93.63
	II	40	83-105	81.7-100.5	90.58	89.11-92.07
	III	27	79-101	77.9-101.2	88.75	86.55-91.02
	IV	113	79-105	80.0-100.8	89.80	88.83-90.78
	V	32	83-105	81.1-101.6	90.77	88.98-92.59
	VI	83	81-102	81.3-100.9	90.59	89.52-91.67
Red blood cells in millions/mm^3	I	126	3.81-5.83	4.13-5.58	4.80	4.74-4.82
	II	40	4.11-5.57	4.12-5.27	4.66	4.57-4.76
	III	27	3.80-5.38	3.81-5.33	4.51	4.36-4.66
	IV	110	3.80-5.57	3.98-5.28	4.58	4.52-4.64
	V	30	4.13-5.52	3.99-5.35	4.62	4.50-4.75
	VI	80	3.79-5.63	4.00-5.38	4.64	4.56-4.72
Hematocrit in %	I	83	35-58	37.9-54 5	45.47	44.57-46.38
	II	23	38-55	37.2-55.1	45.28	43.46-47.18
	III	8	39-46	38.3-49.2	43.44	41.57-45.41
	IV	66	38-55	38.0-52.8	44.74	43.85-45.66
	V	19	40-51	38.9-52.0	44.95	43.48-46.47
	VI	77	34-53	37.8-50.5	43.73	43.02-44.46
MCV in μ^3	I	83	76-115	81.2-105.8	92.65	91.31-94.00
	II	23	85-107	80.9-111.4	94.99	91.88-98.21
	III	8	85-96	82.6-99.4	90.63	87.71-93.65
	IV	66	84-111	83.0-108.0	94.69	93.17-96.22
	V	19	86-106	84.8-107.9	95.64	93.03-98.33
	VI	77	80-114	82.9-106.8	94.07	92.73-95.44

and not significantly different between groups. Group 4 contains one high glucose level of a known diabetic.

Blood urea nitrogen, with a normal range of 10-20 mg% for this method, shows a normal range in all groups considered. Uric acid shows a rather wide normal range: from 2.5-8.0 mg%. Again, all means and ranges are normal, the control group being significantly higher than the other groups. These results may not quite agree with those of Long et al. (1969), who found a significant correlation between increased uric acid levels in the blood and the amount of aldrin handled by the farm workers they examined.

As regards cholesterol all groups show about the same means and ranges within normal limits: 150-300 mg% for this method.

Total bilirubin is normal up to 1.0 mg%. Insecticide workers show normal means and ranges, whereas the control group was influenced by 3 increased individual readings, the highest of which was 2.1 mg%.

Alkaline phosphatase, LDH and SGOT have normal ranges of 30-85, 90-200 and 10-50 mU/ml respectively with the method used. For these parameters means and ranges are normal and there are no significant differences between the groups.

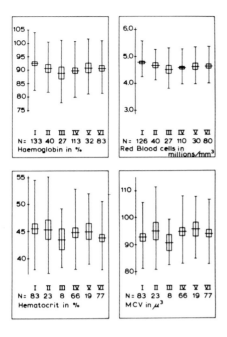

Fig. 19. Geometric mean and 95% confidence ranges of this mean and of all observations of the parameters haemoglobin, red bloodcells, hematocrit and MCV.

It is suggested, as illustrated by the total bilirubin values that the control group may not be considered quite normal in all respects. The members belong to another group of 4-shift workers, who are not occupationally exposed to insecticides, but work with different chemicals. This, together with differences in demographic background and with the fact that blood sampling was not done under quite standardized conditions suggests that occasional small differences found between the means and the confidence limits of some of the groups may not be considered as toxicologically meaningful and the pattern of these slight variations may well be explained by the variations in composition between the groups.

Under the conditions of this study it is concluded that the means and ranges of all 20 parameters examined are normal.

This conclusion is put into proper perspective when it is related to the length and the intensity of exposure to insecticides.

The mean duration of the insecticide exposure in the group of insecticide workers (group IV) is 72.9 months, or a little over 6 years. In the long-term exposure group V the mean exposure even is 159.7 months or $13^{1}/_{3}$ year. Thus group V has more than double the length of exposure of group IV. Again, comparing the parameters of group IV and V, all figures show that there is no effect of this difference in duration of exposure. The possibility that progressive selection might influence the results in these groups has been discussed, and largely dismissed, in chapter 11.

Table 49

WHITE BLOOD CELLS, SEDIMENTATION RATE OF ERYTHROCYTES, PLASMA
CHOLINESTERASE ACTIVITY, SGPT

Parameter	Group	No.	Range	95% conf. range	Geom. mean	95% conf. range of geom. mean
			All observations			
White blood cells	I	133	3.800-11.000	4.175-10.724	6.699	6.432-6.979
per mm^3	II	40	4.000-11.800	3.618-12.097	6.616	6.014-7.278
	III	27	3.600-10.700	3.649-11.158	6.381	5.730-7.106
	IV	113	3.600-11.800	3.913-11.117	6.595	6.282-6.924
	V	32	4.200-11.800	3.970-12.071	6.923	6.275-7.638
	VI	83	4.200-13.300	4.057-11.552	6.845	6.463-7.250
SRE mm 1st hour	I	133	1-29	0.7-17.6	3.6	3.1-4.1
	II	40	1-14	1.1-11.1	3.5	2.9-4.2
	III	27	1-10	1.1-11.4	3.5	2.8-4.4
	IV	86	1-27	0.9-4.5	3.7	3.2-4.3
	V	32	1-15	0.7-15.3	3.3	2.5-4.4
	VI	83	1-23	0.7-15.7	3.4	2.9-4.0
ChE plasma	I	131	52-170	66.2-156.0	101.6	97.9-105.5
Δph/hour x 100	II	39	70-138	74.1-139.0	101.5	96.5-106.8
	III	27	64-150	67.8-159.8	104.1	95.7-113.2
	IV	113	63-150	67.8-143.5	98.6	95.2-102.2
	V	30	66-130	72.7-136.8	99.7	94.1-105.7
	VI	70	60-136	64.6-151.3	98.9	94.0-104.0
SGPT in K.U.	I	130	5-64	5.3-28.3	12.2	11.3-13.1
	II	39	6-28	5.9-23.8	11.8	10.5-13.2
	III	26	6-19	7.9-24.2	13.8	12.4-15.4
	IV	112	4-28	5.3-25.2	11.5	10.7-12.4
	V	30	5-22	5.3-22.1	10.9	9.6-12.4
	VI	71	3-24	4.1-25.3	10.2	9.2-11.4

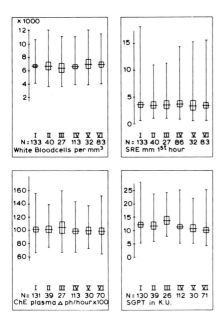

Fig. 20. Geometric mean and 95% confidence ranges of this mean and of all observations of the parameters white bloodcells, sedimentation rate of erythrocytes, plasma cholinesterase and SGPT.

The intensity of insecticide exposure is reflected in the insecticide levels in the blood as tabulated in Table 47 and illustrated in Fig. 18. As already mentioned, endrin was not detected in any of the workers at a detection threshold of 0.005 μg/ml. In the group of the non-insecticide workers dieldrin and Telodrin were not detected in the blood of any of the workers at a threshold level of 0.005 and 0.002 μg/ml respectively.

As to the other groups examined, the only significant differences between the groups are that dieldrin and Telodrin, and thus also the D.Eq., are lower in the group of endrin operators (group III); dieldrin is also lower in the group of the former workers (VI), which was to be expected. The fact that Telodrin levels in this latter group are still high is largely due to the fact that this group contains many former Telodrin workers who were transferred to other work on account of Telodrin intoxication or Telodrin levels in the blood exceeding safety limits, and of course also to the long half-life of Telodrin in the blood.

Means and top-levels of insecticides (top-levels in brackets) in the blood of all insecticide workers and of those with exposures longer than 12 years (groups IV and V respectively) are given below.

Table 50

CALCIUM, INORGANIC PHOSPHATE, GLUCOSE AND BLOOD UREA NITROGEN
IN BLOOD SERUM

| Parameter | Group | All observations | | | | 95% conf. range of geom.mean |
		No.	Range	95% conf. range	Geom. mean	
Calcium^{++} in mg%	I	129	9.5-11.6	9.4-10.9	10.14	10.07-10.21
	II	40	9.2-11.5	9.0-11.1	9.99	9.83-10.15
	III	26	9.0-10.4	9.1-10.6	9.81	9.67-9.95
	IV	113	9.2-11.5	9.1-10.9	9.93	9.84-10.02
	V	32	9.2-10.8	9.1-10.7	9.87	9.74-10.01
	VI	82	9.1-10.9	9.0-10.7	9.83	9.74-9.92
Inorganic phosphate in mg%P	I	131	1.7-4.8	2.25-4.77	3.27	3.17-3.38
	II	40	2.5-4.5	2.62-4.80	3.57	3.39-3.75
	III	26	2.3-4.4	2.69-4.67	3.54	3.35-3.74
	IV	113	2.3-5.0	2.62-4.72	3.53	3.44-3.64
	V	32	2.6-5.0	2.59-4.77	3.52	3.33-3.71
	VI	83	2.4-4.7	2.58-4.65	3.46	3.35-3.58
Glucose in mg%	I	129	55-135	61.5-129.2	89.2	86.3-92.1
	II	40	50-135	57.7-141.2	90.2	84.1-96.8
	III	26	65-115	70.1-121.0	92.1	87.3-97.2
	IV	113	50-270	61.4-146.8	94.9	91.1-98.9
	V	31	50-130	62.2-140.5	93.5	86.9-100.6
	VI	82	60-160	62.2-147.3	95.7	91.2-100.4
BUN in mg%	I	129	8-26	9.9-23.9	15.37	14.78-15.98
	II	40	10-22	9.8-21.2	14.41	13.55-15.32
	III	26	10-22	9.5-25.5	15.59	14.16-17.16
	IV	113	10-27	10.0-23.1	15.20	14.62-15.81
	V	32	10-27	9.2-23.8	14.79	13.60-16.09
	VI	81	9-28	10.1-24.9	15.87	15.10-16.69

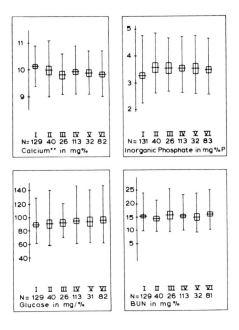

Fig. 21. Geometric mean and 95% confidence ranges of this mean and of all observations of the parameters Calcium[++], inorganic phosphate, glucose and blood urea nitrogen.

Insecticide levels in whole blood in µg/ml: means and top-levels

	Dieldrin	*Telodrin*	*D. Eq.*
Group IV: all insect. workers	0.0085 (0.149)	0.00172 (0.008)	0.0173 (0.150)
Group V: long-term exposure	0.0081 (0.060)	0.00251 (0.008)	0.0230 (0.120)

If only dieldrin is considered, the mean blood levels of 0.0085 and 0.0081 µg/ml correspond, according to the formula of Hunter and Robinson (1969), to at least an average equivalent oral daily intake of 99 and 94 µg/man/day, which compared to the latest calculations of the average daily intake of the general population in the U.K. and the U.S.A. of 7 µg/man/day, is 14 times and 13 times that intake of this general population.

This calculation for the top-levels of dieldrin alone of 0.149 and 0.060 µg/ml respectively gives average daily intakes of 1733 and 698 µg/man/day, or 248 and 100 times the intake of the general population in the U.K. and the U.S.A.

Up till now we have not taken into account the fact that the total of dieldrin and Telodrin expressed as D.Eq. is approximately 2½ times the value for dieldrin alone, nor have we considered the effect of the biologically more active endrin (higher toxicity and more rapid metabolism) which does not find reflection in a measurable blood level. The average daily intake of endrin and Telodrin in the general population is zero.

Table 51

URIC ACID, CHOLESTEROL, TOTAL PROTEIN AND ALBUMIN IN BLOOD SERUM

Parameter	Group	All observations				95% conf. range of geom. mean
		No.	Range	95% conf. range	Geom. mean	
Uric acid in mg%	I	130	3.0-8.9	4.3-8.5	6.01	5.83-6.20
	II	40	2.9-8.3	3.7-8.3	5.50	5.16-5.87
	III	26	3.9-8.8	3.6-7.9	5.38	4.99-5.81
	IV	113	2.9-8.8	3.7-8.0	5.42	5.22-5.62
	V	32	3.7-7.4	3.9-7.6	5.41	5.10-5.75
	VI	81	3.4-7.7	4.0-7.6	5.53	5.34-5.73
Cholesterol in mg%	I	125	130-380	156-339	230.3	222.5-238.4
	II	39	160-385	100-350	236.9	222.5-252.2
	III	26	170-315	170-288	221.1	210.0-232.8
	IV	113	140-405	159-338	231.5	223.4-239.8
	V	32	175-355	175-332	241.0	227.7-255.1
	VI	79	160-370	169-317	231.8	223.7-240.1
Total protein in g%	I	130	6.6-8.3	6.8-8.1	7.44	7.39-7.50
	II	40	6.3-8.0	6.7-8.0	7.31	7.20-7.42
	III	26	6.4-7.8	6.5-7.9	7.14	7.00-7.28
	IV	113	6.3-8.3	6.5-8.0	7.21	7.14-7.29
	V	32	6.4-7.0	6.5-7.8	7.15	7.04-7.26
	VI	83	6.3-8.1	6.6-7.9	7.25	7.18-7.33
Albumin in g%	I	129	3.2-5.3	3.6-5.3	4.39	4.32-4.47
	II	40	3.3-4.9	3.5-5.1	4.20	4.07-4.32
	III	26	3.3-4.7	3.4-4.9	4.06	3.91-4.22
	IV	113	3.2-5.1	3.4-5.1	4.14	4.06-4.22
	V	32	3.3-4.8	3.3-5.1	4.09	3.93-4.25
	VI	83	3.5-5.0	3.6-5.1	4.33	4.24-4.41

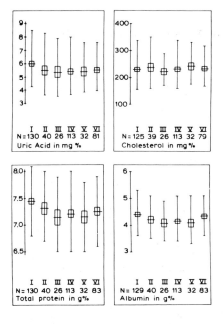

Fig. 22. Geometric mean and 95% confidence ranges of this mean and of all observations of the parameters uric acid, cholesterol, total protein and albumin in serum.

Moreover, we know from daily practice over the last ten years, as well as from the follow-ups of insecticide levels since 1964, that insecticide exposures in our workers must have been much higher during earlier production years. See for instance Table 20 in chapter 13. Thus in the long-term exposure group the total accumulated intake must have been much higher than corresponds with the present concentration in the blood.

19.5   SUMMARY

Results of 20 biological parameters: Hb, RBC's, hematocrit, MCV, WBC's, SRE, plasma ChE, SGPT, Ca^{++}, inorg. phosphate, glucose, BUN, uric acid, cholesterol, total protein, albumin, total bilirubin, alkaline phosphatase, LDH and SGOT were compared in a group of insecticide workers, a group of former insecticide workers and a group of oil cracking and plastics operators.

All results were within normal limits, and those slight differences which were found between groups, could be attributed to variations in demographic background between the groups examined.

The group of insecticide workers consisted of 113 workers with a mean duration of insecticide exposure of over 6 years and a mean calculated daily intake of dieldrin (Telodrin and endrin not taken into account) of at least 14 times that of the general population in the U.S.A. and the U.K. The top-levels in this group correspond to an average daily intake of 248 times that of this general population.

Table 52

TOTAL BILIRUBIN, ALKALINE PHOSPHATASE, LDH AND SGOT IN BLOOD SERUM

Parameter	Group	All observations				95% conf. range of geom.mean
		No.	Range	95% conf. range	Geom. mean	
Total bilirubin in	I	129	0.1-2.1	0.14-1.41	0.45	0.41-0.49
mg%	II	40	0.1-1.0	0.14-0.93	0.36	0.31-0.42
	III	26	0.1-0.7	0.13-0.81	0.32	0.27-0.39
	IV	113	0.1-1.0	0.09-0.93	0.29	0.26-0.32
	V	32	0.1-0.8	0.09-1.02	0.30	0.24-0.37
	VI	83	0.1-0.8	0.10-1.25	0.35	0.30-0.40
Alkaline phosphatase	I	131	15-90	27.1-91.4	49.8	47.2-52.5
in mU/ml	II	40	35-75	32.5-79.0	50.7	47.2-54.3
	III	26	35-90	34.4-90.0	55.7	50.7-61.2
	IV	113	25-115	32.4-91.9	54.5	51.9-57.3
	V	32	35-85	36.2-89.4	56.9	52.5-61.6
	VI	83	25-85	29.5-90.1	51.5	48.5-54.8
LDH in mU/ml	I	125	105-235	123-226	166.5	162.0-171.2
	II	40	110-260	117-227	163.2	154.8-172.0
	III	26	130-200	128-206	162.2	154.8-170.0
	IV	113	110-260	120-226	164.8	159.9-169.7
	V	32	125-200	127-216	166.0	158.4-173.9
	VI	83	100-225	108-235	159.6	152.9-166.5
SGOT in mU/ml	I	131	20-140	22-74	40.2	38.1-42.4
	II	40	15-65	17-74	35.9	32.0-40.3
	III	26	15-60	21-72	38.2	33.8-43.2
	IV	113	15-65	20-68	36.5	34.4-38.7
	V	32	15-60	17-71	34.4	30.3-39.1
	VI	83	15-70	20-70	37.5	35.0-40.2

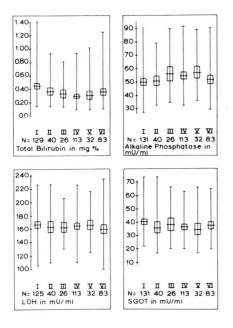

Fig. 23. Geometric mean and 95% confidence ranges of this mean and of all observations of the parameters total bilirubin, alkaline phosphatase, LDH and SGOT in serum.

A subgroup of 32 insecticide workers with a mean insecticide exposure of $13\frac{1}{3}$ years at the time of the last examination had a calculated daily intake of dieldrin alone of at least 13 times that of the general population: whereas top-levels in this group correspond to an average daily intake of 100 times that of the general population. Also, in these calculations endrin and Telodrin exposures have not yet been taken into account. This would more than double the ratio.

Exposures of these groups in previous years are known to have been much higher than the calculated intakes corresponding to blood levels at this time.

Summarizing we may conclude from this part of the study that exposure for more than 12 years to dieldrin together with related insecticides, with dieldrin intakes of at least 13 times those of the general population, did not show any effects in the 20 parameters measured.

Also in the group of former insecticide workers, which includes many workers who previously had an insecticide intoxication and many workers who were transferred on account of blood levels exceeding our safety levels, no effect on these 20 parameters was apparent.

Chapter 20

Study of parameters of enzyme induction of all insecticide workers

20.1 INTRODUCTION

In chapter 2.1.5 the phenomenon of enzyme induction has been discussed at some length. Enzyme induction is the enhancement of the activity of those microsomal enzyme systems which detoxify some, mostly lipid soluble, drugs and chemicals. This may result in an enhanced metabolism of other drugs and chemicals or body substrates, and thus might interfere with drug therapy or even hormonal regulations in the body.

Enzyme induction is in many cases the first measurable response of the body to continuous low-dose exposure to drugs and chemicals. It may or may not be a harmful effect. A certain degree of enhanced detoxification might often even be beneficial to the organism. Although this is still a controversial topic, two things should be kept in mind:
1. A daily dose exists below which the effect of enzyme induction is no longer measurable.
2. The effect is reversible.

In chapter 2.5.5 various tests for enzyme induction were discussed, most of which were unsuitable for routine use in preventive medicine.

For our study of enzyme induction in occupationally exposed insecticide workers we selected:
a. determination of serum enzyme levels,
b. determination of pp'DDE levels in whole blood, and
c. determination of the ratio of 6-β-hydroxycortisol and 17-hydroxycorticosteroids excretion in urine.

The results will be discussed in that order.

20.2 SERUM ENZYME LEVELS

In chapters 18 and 19 results have been evaluated of the determination of serum alkaline phosphatase, SGOT, SGPT and serum-LDH levels respectively, in various blood insecticide level groups and in all insecticide workers as compared with those of a control group. In addition plasma cholinesterase activities were discussed in chapter 19.

Serum alkaline phosphatase levels were within normal limits in all groups. However there was a statistically significant increase in the 0.21-0.30 D.Eq. exposure group. No significant differences were found between the various exposure groups in chapter 19.

SGOT levels also were within normal ranges and showed normal means in all exposure groups in both studies, and there was no difference when compared with a control group. SGOT levels were significantly lower in the 0.01-0.05 D.Eq. group compared with those of higher exposure as well as those of the <0.01 D.Eq. group. With the Spearman rank correlation tests a very slight, but statistically significant increase of SGOT with increasing D.Eq. was found.

Levels of *SGPT* again had normal means and ranges in all groups. There was no difference with the control group. SGPT in the endrin group (III) was slightly, though not significantly, higher than in the other exposure groups. There was no significant difference between the D.Eq. exposure groups and with the Spearman rank correlation test an increase of SGPT with increasing D.Eq. was only just demonstrable.

Levels of *serum-LDH* showed normal ranges and means in all groups. There were no significant differences between groups and with the control group. No significant correlation between D.Eq. blood level and serum LDH was found.

Plasma cholinesterase activity was compared only in the groups in chapter 19. Means and ranges were normal in all groups and there were no differences between groups.

The fact that SGOT, SGPT and LDH showed normal patterns in all groups means that the integrity of the liver cell membrane, and thus of the liver, was intact in all groups. As was discussed in chapter 18, the finding of an increased serum alkaline phosphatase in people with a blood insecticide concentration exceeding our safety threshold level, *i.e.* in the 0.21-0.30 D.Eq. exposure group should then be regarded as a sign of adaptation of the liver cell and thus may be an indirect sign of enzyme induction.

20.3   pp'DDE LEVELS IN WHOLE BLOOD

As was discussed in chapter 2.5.5 pp'DDE levels in blood are regarded as a reliable test for DDT exposure in the general population. It was mentioned that some known enzyme inducers like phenobarbitone and diphenylhydantoin reduced pp'DDE blood levels (Davies *et al.* 1969-a). It is also known, that in chronic feeding studies with dieldrin the administration of DDT causes an increased excretion of dieldrin metabolites resulting in lower dieldrin levels in the body, compared with those of control animals not receiving DDT (Street *et al.* 1967).

For these reasons it was felt that pp'DDE levels in blood might be a good indicator for enzyme induction and we started determining pp'DDE levels in the blood together with the other insecticide levels. As DDT was not, and is not, handled in our plants, all our workers should exhibit pp'DDE levels in the blood of the same order as those of the general population.

Together with the determination of the parameters and insecticide levels described in chapter 19, pp'DDE levels were examined in those same groups of workers. In addition there was a second non-insecticide exposed control group, consisting mainly of refinery workers. Both control groups, the one just mentioned and the plastics and cracking operators, may be regarded as representing the general population of Holland as far as pp'DDE levels in the blood are concerned.

The results of this exercise are tabulated in Table 53 and illustrated in Fig. 24. It soon became apparent, that lower pp'DDE levels were found in our insecticide workers, and on closer observation it appeared that only endrin workers and former endrin workers showed this trend.

From Table 53 and Fig. 24 it is apparent that:

a. The mean pp'DDE level in whole blood is of the same order in both control groups: 0.0132 and 0.0127 μg/ml respectively. Ranges also were similar.

b. Insecticide workers who handle (or have handled) aldrin, dieldrin (and Telodrin) but no endrin have pp'DDE levels in blood, the ranges and means of which do not differ significantly from those of the control groups. When we select from this group of aldrin-dieldrin workers the 10 with the highest dieldrin levels in the blood, these levels range from 0.080-0.150 with an arithmetic mean of 0.1053 μg/ml. In these 10 samples pp'DDE levels range from 0.009-0.034 with an arithmetic mean of 0.0144 μg/ml; thus of the same order as those of the control groups. In other words even at this estimated average daily intake of 1224 μg/man/day dieldrin, which is about 175 times the intake of the general population, no effect on pp'DDE levels in the blood occurred.

c. In endrin workers and former endrin workers significantly lower pp'DDE levels were found in the blood, with both significantly lower means and lower top-levels. Only in endrin workers we did find pp'DDE levels below the detection level of 0.005 μg/ml. 15

Table 53

pp'DDE LEVELS IN WHOLE BLOOD in μg/ml

Group	No.	Geom. mean	95% conf. range of geom. mean	Range of all observations	95% conf. range of all observations
Control group 1: plastic and cracking operators	130	0.0132	0.0122 - 0.0143	0.005 - 0.058	0.0053 - 0.0333
Control group 2: refinery operators	80	0.0127	0.0115 - 0.0141	0.005 - 0.052	0.0051 - 0.0317
Aldrin-dieldrin workers	63	0.0124	0.0110 - 0.0138	0.005 - 0.042	0.0051 - 0.0302
Former aldrin-dieldrin workers	42	0.0130	0.0112 - 0.0150	0.005 - 0.031	0.0049 - 0.0339
Endrin workers	29	0.0045	0.0034 - 0.0059	<0.005 - 0.018	0.0010 - 0.0197
Insect. workers former endrin	23	0.0051	0.0039 - 0.0066	<0.005 - 0.018	0.0015 - 0.0179
Former insecticide former endrin	39	0.0062	0.0051 - 0.0077	<0.005 - 0.027	0.0017 - 0.0229

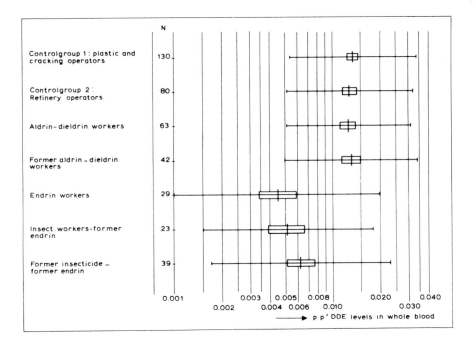

Fig. 24. p.p'DDE levels in whole blood in μg/ml. (Geometric mean and 95% confidence limits of this mean and of all observations).

out of 29 endrin workers had pp'DDE levels below this detection level.

Former endrin workers, now working in aldrin-dieldrin or in non-insecticide plants showed somewhat higher means and ranges of pp'DDE levels, but had not yet reversed to the same range as the general population; even in those workers who had been away from the endrin plant for longer than 5 years, this was still a remarkable feature.

Summarizing from the results of pp'DDE determination in blood we may conclude that aldrin and dieldrin in exposures up to at least 175 times the exposure of the general population do not stimulate the enzyme systems which metabolize pp'DDE. Occupational exposure in endrin manufacturing however clearly induces this enzyme activity.

20.4   6-β-HYDROXYCORTISOL EXCRETION IN URINE

As was mentioned in chapter 2.1.5 and 2.5.5, many drugs and chemicals may stimulate the hydroxylation of steroids in the body, amongst them phenobarbital, diphenylhydantoin, phenylbutazone and N-phenylbarbital (Werk et al. 1964, Burstein et al. 1965, Kuntzman et al. 1966, Conney 1967, Kuntzman et al. 1968).

The studies of these investigators suggest that the measurement of the urinary excretion of 6-β-hydroxycortisol, a metabolite of cortisol, compared with the excretion of total 17-hydroxycorticosteroids, which is not changed by the inducers, might be a useful index for the induction of hydroxylase in liver microsomes in man. Kuntzman *et al.* (1968) found that 6-β-OH-cortisol excretion in man is normally below 400 µg/day, whereas in situations of enzyme induction, such as in N-phenylbarbital treated human volunteers, excretions exceeding 400 µg/day were found.

For this study excretion of 6-β-OH-cortisol and 17-OH-corticosteroids were determined in 20 non-insecticide exposed four-shift workers, as well as in 13 four-shift aldrin-dieldrin workers and 8 four-shift endrin workers. All urine samples were collected between 8 a.m. and 11 a.m. on the last day of the morning shift. Hormone determinations were made by Searle Scientific Services*, who also added their own control group of 10 men. By determining the ratio between the excretion of 6-β-hydroxycortisol and 17-OH-corticosteroids, the factor diuresis is eliminated and there is no need for examining 24-hour urine samples. Therefore this method is convenient for use in healthy workers.

The results of these determinations are given in Table 54 and 55 and in Fig. 25. From this it is clear that geometric means and ranges of the ratio in aldrin-dieldrin workers does not differ from those in the control groups. In this group of 13 aldrin-dieldrin workers the range of pp'DDE in the blood was 0.006-0.042 µg/ml with an

Table 54

$$\text{RATIO} \frac{6-\beta-\text{OH}-\text{CORTISOL IN } \mu\text{g/L}}{17-\text{OH}-\text{CORTICOSTEROIDS IN mg/L}} \text{ IN URINE}$$

Group	No.	Geom. mean	95% conf. range of geom. mean	Range of all observations	95% conf. range of all observations
Control group (Searle)	10	27.48	18.70 - 40.38	12.3 - 58.2	8.13 - 92.84
Control group (Shell)	20	26.58	20.13 - 35.11	6.7 - 82.1	7.66 - 92.24
Aldrin-dieldrin workers	13	25.32	16.66 - 38.48	5.9 - 61.8	5.59 - 114.55
Endrin workers	8	87.28	67.22 - 113.32	59.1 - 125.9	41.70 - 182.68

arithmetic mean of 0.015 µg/ml, thus in the same range as in the general population (see chapter 20.3). Dieldrin levels in the blood of these 13 workers ranged from 0.018-0.110 µg/ml with an arithmetic mean of 0.051 µg/ml. In these workers, who had reached a steady state level as far as dieldrin is concerned, the mean level corresponds, according to the formula of Hunter and Robinson (1969), to an average equivalent oral daily intake of 593 µg/man/day, which is at least 85 times the intake of the general population in the U.K. and the U.S.A. for this insecticide. But even the man with the highest dieldrin level

---

*Searle Scientific Services, Lane End Road, High Wycombe, Bucks., U.K.

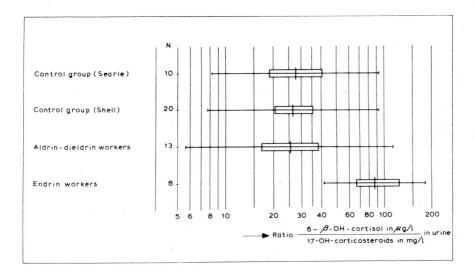

Fig. 25. Ratio $\dfrac{\text{6-}\beta\text{-OH-cortisol in }\mu g/L}{\text{17-OH-corticosteroids in mg/L}}$ in urine. (Geometric mean and 95% confidence limits of this mean and of all observations).

in this group – 0.110 μg/ml, or 183 times the present blood level of the general population – showed, apart from a pp'DDE level of 0.014 μg/ml, a 6-β-hydroxycortisol ratio of $\frac{166}{5.6}$ = 29.6, both values being quite normal in comparison with the control groups.

The ratio between the excretion of 6-β-OH-cortisol and 17-OH-corticosteroids was also determined in 8 endrin workers. There we found a range and a geometric mean of this ratio, which was significantly higher when compared with both the control groups and the aldrin-dieldrin workers. The 6-β-OH-cortisol levels in urine were higher in this group, whereas 17-OH-corticosteroids and 17-ketosteroids did not show marked differences between groups. Of course these levels expressed in mg/l are dependent on the total urine production per day. All hormone levels found in urine of all groups are tabulated in Table 55.

20.5   DISCUSSION

Of the serum enzyme levels examined, alkaline phosphatase showed a statistically significant increase. It is emphasized, however, that the means and ranges were all still normal even in those workers with a blood D.Eq. above our intoxication threshold. This statistically significant increase may have been caused by a certain degree of interference with alkaline phosphatase excretion due to hepatocyte enlargement, or perhaps also to a certain degree of induction of alkaline phosphatase synthesis in the liver.

The fact that a rank correlation test showed an increase of the means of SGOT and SGPT with increasing D.Eq. – although all values were normal – is not in contradiction

Table 55

HORMONE EXCRETION IN URINE

Group	6 - β - OH - cortisol in µg/L	17 - OH - corticosteroids in mg/L	17 - keto - steroids in mg/L	Ratio $\frac{6 - \beta - OH - c.}{17 - OH - c.}$
Control group	540	27.6	26.1	19.6
(Shell)	540	18.2	25.8	29.7
	1250	20.3	27.0	61.6
	370	17.0	14.6	21.8
	340	12.4	21.2	27.4
	220	9.5	10.7	23.2
	72	10.7	14.5	6.7
	420	11.8	15.1	35.6
	690	11.3	12.4	61.1
	894	17.8	21.2	50.2
	40	2.6	5.1	15.4
	96	11.7	16.6	82.1
	268	7.4	17.5	36.2
	224	12.0	23.0	18.7
	250	8.6	9.6	29.1
	152	7.0	7.0	21.7
	138	9.4	12.1	14.7
	144	9.7	7.6	14.8
	178	9.3	9.1	19.2
	416	11.0	18.0	37.8
Aldrin-	108	6.0	14.9	18.0
dieldrin	520	11.3	18.4	46.0
workers	740	20.2	26.7	36.6
	54	3.9	2.1	13.8
	128	4.2	7.0	30.5
	166	5.6	6.0	29.6
	80	3.4	3.5	23.5
	76	5.6	7.3	13.6
	990	21.0	26.8	47.1
	92	6.4	6.3	14.4
	588	10.0	12.0	58.8
	766	12.4	15.8	61.8
	72	12.2	18.1	5.9
Endrin workers	1510	17.5	18.1	86.3
	1740	15.9	15.3	109.8
	340	2.7	1.7	125.9
	1380	14.6	26.8	94.5
	1040	16.4	21.4	63.4
	668	11.3	15.1	59.1
	1160	18.3	30.4	63.4
	1445	11.5	13.5	125.7

with this finding. It only proves, that at these high blood levels the integrity of the liver cell is intact and the function of the liver cell membrane may get stressed, but is not disturbed.

In our aldrin-dieldrin workers no indication was found of enzyme induction, either with pp'DDE determination in the blood or with the determination of 6-β-OH-cortisol excretion in the urine. With the first method we were able to show that even a calculated average daily intake of dieldrin of 175 times that of the general population did not influence average pp'DDE levels in the blood. With the second method, calculated average daily intakes of dieldrin of 85 times and a top-level of 183 times that of the general population did not influence steroid metabolism in man, *i.e.* enhanced cortisol metabolism was not apparent in an increase of the ratio 6-β-OH-cortisol/17-OH-corticosteroids.

That both methods are suitable to demonstrate the occurrence of enzyme induction is illustrated by our endrin workers. pp'DDE blood levels, both means and ranges, were significantly lower in the group of endrin workers and former endrin workers as compared with the control groups and the aldrin-dieldrin workers. In endrin workers the urine samples examined showed a significant increase of 6-β-OH-cortisol excretion in the ratio of 6-β-OH-cortisol/17-OH-corticosteroids when compared with controls and with aldrin-dieldrin workers. Both results should be regarded as signs of enzyme induction. The problem with endrin, however, is that up to now we have been unable to quantify the exposure. In our opinion, endrin exposure in our endrin workers is of the same magnitude as the dieldrin exposure in our dieldrin workers. As endrin is rapidly metabolized in the body, we do not normally find endrin in the blood of our workers at our detection threshold of 0.005 μg/ml. Our monitoring system for endrin, however, has to be improved either by lowering our detection threshold until we are able to measure exposure, or by the determination of endrin metabolites in urine.

At this time we can only conclude that enzyme induction is found in our endrin workers. Although this might be caused by endrin, there is no proof of this in our results. Another chemical used or one of the intermediates, *e.g.* isodrin, might in fact be the causative agent.

As far as the general population is concerned, this finding of enzyme induction is hardly relevant as, in normal circumstances, endrin is never detected in food, drinking water or air (see chapter 7.3). Also in those areas where endrin is most extensively used, as for instance the lower Mississippi delta, pp'DDE levels in fat do not differ from those in other areas (Hayes 1965).

Richardson *et al.* (1967) fed 3 beagles a daily dose of 0.1 mg/kg (equivalent to approximately 4 ppm) endrin during 4 months. Stationary endrin levels in the blood were reached of 0.0020-0.0084 μg/ml. DDE and DDT levels in the blood remained at essentially the same levels of 0.002 μg/ml throughout the experiment.

There are at least three possible explanations for the discrepancy between these results and our findings in man. Endrin metabolism might be either more rapid in man (so that higher daily doses are needed for the same blood level to be reached as compared to the dog), or dogs might be less inclined to react with enzyme induction, or one of the intermediates or other chemicals used in endrin manufacture and not endrin, might be the causative agent.

At this time it is only possible to mention the enzyme induction found in our endrin

204

workers, and also, the discrepancy between the established rapid detoxification of endrin in the body and the long-lasting effect of enzyme induction found in our former endrin workers. It is evident that more work has to be done before these phenomena have been fully analyzed and understood.

20.6 SUMMARY

Plasma cholinesterase activity did not show any influence of organochlorine exposure in our exposure groups.

Serum alkaline phosphatase, SGOT and SGPT may be indicators for enzyme induction at high exposure levels. Compared with pp'DDE levels in blood and 6-β-OH-cortisol excretion in urine they are not early indicators of this phenomenon.

In aldrin-dieldrin workers no indication of enzyme induction was found with the methods used. pp'DDE levels showed no effect with exposures of at least 175 times that of the general population; 6-β-OH-cortisol excretion in urine as a measure for steroid metabolism in the body remained normal with exposures of at least 183 times those of the general population.

In our endrin workers enzyme induction was shown to occur with both methods. More work has to be done to study this response relationship. If endrin is proved to be the causative agent, this finding is still irrelevant to the general population, since endrin is not normally found in food, air and water.

F. DISCUSSION AND SUMMARY

Chapter 21

Discussion of results; conclusions

21.1   STUDY OF LONG-TERM OCCUPATIONAL EXPOSURE GROUP

In chapters 12-15 inclusive, a review was given of medical data of the long-term exposure group of all 233 workers with occupational exposure of more than 4 years. This group was selected from the total of 826 workers who had been employed in organochlorine insecticide manufacturing over the past 15 years prior to January 1st 1968.

A separate study was made of extreme exposure groups within this long-term exposure group: those with exposures longer than 10 years (n = 35), those with blood concentrations above safety threshold levels (n = 17) and those with clinical intoxications (n = 9).

Our medical selection policy for the insecticide workers was discussed in chapter 11 and the conclusion was that this group of insecticide workers is not essentially different from any other group from the general working population.

Apart from temporary signs and symptoms of specific intoxication in some of the workers, no findings which might be regarded abnormal in a group of this size, and which could in any way be related to insecticide exposure, were found on studying the medical files of these workers. There was no deterioration in the course of any pre-existent disease.

Those workers who were transferred on account of insecticide intoxications or insecticide levels in blood exceeding our safety threshold level, did not afterwards have an abnormal disease pattern.

In the routine medical examination of this whole group of 233 workers no unexpected abnormalities for a group of this size were found in all parameters examined, which included body weight, blood pressure, haemoglobin, white blood cells, sedimentation rate of erythrocytes, serum alkaline phosphatase, SGOT, SGPT, LDH, total serum protein and serum protein spectrum. We found no indication of permanent injury to the body, particularly no disturbances of the central nervous system or the blood and no hepatic or renal damage.

Although this part of the study included data up to January 1st 1968, the additional two years exposure since that time have made no difference in the favourable health condition described above.

Those pre-exposure parameters available from extreme exposure groups and a suitable control group: *i.e.* body weight, blood pressure, haemoglobin, white blood cell count and sedimentation rate of erythrocytes, were re-examined 10 years later. There was no significant difference between the results of the extreme exposure groups and the control group.

The significance of these results becomes clear when we consider the severity of the insecticide exposure in these groups. Insecticide blood levels for the groups under discussion were tabulated in Tables 18 and 19. Although blood levels could not be measured in earlier production years, other medical and technical observations make it clear that in those preceding years the insecticide exposure of our insecticide workers was at least similar, and most probably higher, than in subsequent years.

If dieldrin levels only are taken into account, the average dieldrin blood level of the total long-term exposure group (n = 233) over the 4 years considered is 0.035 μg/ml, which, according to the formula of Hunter and Robinson (1969) corresponds to an approximate average equivalent oral daily intake of 407 μg/man/day of dieldrin, which, according to the data discussed in chapter 5.7, corresponds to 58 times the present daily intake of dieldrin of the general population in the U.K. and the U.S.A.

When the average of total insecticide exposure of this long-term exposure group, as expressed in the dieldrin equivalent (D.Eq.), is taken, this is 0.120 μg/ml D.Eq. in the blood, which is about 200 times the D.Eq. found in this general population. This still does not take into account the endrin exposure of these workers, which does not normally find an expression in a detectable blood level.

From this part of the study it may be concluded that an occupational exposure of a normal male working population, for 4 to 15 years, at exposures of more than 50 times that of the general population if dieldrin only is considered, or more than 200 times that exposure if all four insecticides are considered — if remaining below the accepted safety levels — does not have an adverse influence on the health of these workers as far as could be detected by all routine examinations and the parameters used therein.

21.2   STUDY OF ALL INSECTICIDE INTOXICATIONS

From a study of all 54 insecticide intoxications which have occurred in our insecticide plant during the earlier production years, it was evident that these intoxications were, without exception, the result of accidental gross over-exposure, or of accumulation as a result of shortcomings in the technical equipment and/or negligence of safety and hygienic precautions, or of a combination of the two.

Three types of organochlorine insecticide intoxication could be distinguished:

Type 1: an acute convulsive intoxication without (or with only very minor prodromi following acute gross over-exposure.

Type 2: a cumulative intoxication as a culmination of regularly repeated smaller doses. Non-specific but consistently pertinent signs and symptoms may be a manifestation of this type of intoxication.

Type 3: an acute convulsive intoxication following an insignificant overdose superimposed on, and triggering off, an accumulative intoxication which was not associated with prodromi, signs or symptoms.

Which of these types of intoxication occurs depends not only on the size and the frequency of the dose, but also on the degree of acute toxicity and the biological half-life of the insecticide.

The classification in these various types of intoxication provides a practical and theoretical understanding, not only of the cases of intoxication which occurred in our insecticide plant, but also of the symptomatology of the cases described in the literature (5.1 and 7.1).

We found no signs or symptoms of insecticide intoxication when insecticide concentrations in the blood were below our safety threshold levels. These, in our experience are:

0.20 µg/ml for dieldrin, which is 333 times the present calculated dieldrin level in the blood of the general population in the U.S.A. and the U.K.,

0.015 µg/ml for Telodrin, and is suggested to be,

0.050-0.100 µg/ml for endrin.

This also is in agreement with results published in the literature.

All cases of insecticide intoxication which occurred in our plant fully recovered to normal: within a few days for endrin, within weeks for aldrin and dieldrin, whereas in Telodrin intoxications this, in some cases, took 6 months.

In all cases of insecticide intoxication associated with typical EEG changes which could be followed up, EEG patterns returned to normal or to a pre-existent atypical pattern. Workers with pre-existent atypical EEG changes were not more prone to insecticide intoxication than those with normal EEG's.

The mean insecticide levels in blood at the moment of transfer were: for all preventive transfers (Table 32) 0.303 D.Eq., or 505 times the D.Eq. of the general population, and for all insecticide intoxications (Table 32) 0.320, or 533 times the D.Eq. of the general population. These data also are in agreement with the sparse blood level data found in medical literature.

From this we may conclude not only that all cases of insecticide intoxication fully recovered to normal, but also that with dieldrin levels in blood of at least 333 times that of the general population in the U.K. and the U.S.A., no signs or symptoms of intoxication occurred. Signs and symptoms of intoxication became apparent only at an average insecticide level in blood, expressed in D.Eq., of more than 500 times that of the general population.

21.3   STUDY OF PARAMETERS OF ALL INSECTICIDE WORKERS

In the chapters 17-20 inclusive as well as in part of chapter 13, various parameters have been discussed.

Absenteeism due to disease and accidents of our insecticide workers over the years 1966-1969 inclusive, as measured by disability rate, frequency rate and severity rate, was approximately equal to that for all other chemical operators, when the age difference between groups was taken into account. Also no disease group occurred significantly more frequently in any of our chemical plants including the insecticide plant.

These results confirm the results of the study of Hoogendam *et al.* (1965) covering an earlier period, and are in agreement with the conclusions of Hayes (1968) from a similar study in the U.S.A.

A study of 8 hematological and biochemical parameters plotted against the dieldrin equivalent measured in the same blood sample revealed that white blood cell count and sedimentation rate of erythrocytes remained unaltered up to a D.Eq. of at least twice our safety threshold level. The same was true for serum LDH and total serum proteins, but for these parameters only data below this threshold level were available. SGOT, SGPT and Hb showed a slight, but with a rank correlation test significant, increase with increasing D.Eq. Serum alkaline phosphatase showed a statistically significant increase in the 0.20-0.30 μg/ml blood D.Eq. group. All ranges and means for all these parameters remained, however, within normal limits.

More recently 20 parameters were examined: haemoglobin, red blood cells, hematocrit, MCV, white blood cell count, sedimentation rate, plasma cholinesterase, SGPT, and a serum chemistry graph consisting of calcium^{++}, inorganic phosphate, glucose, blood urea nitrogen, uric acid, cholesterol, total protein, albumin, total bilirubin, alkaline phosphatase, LDH and SGOT. This was done in groups of insecticide workers, former insecticide workers and non-insecticide workers. This study did not reveal any abnormal results, and those significant differences found between group-means appeared to be the result of differences in demographic composition of the groups and not attributable to insecticide exposure. This becomes clear from comparison of the groups of insecticide workers with exposures of more than 12 years, and former insecticide workers, including many former insecticide intoxications and preventive transfers. No significant differences between these groups were found in all parameters examined.

The mean insecticide level in the blood of 32 insecticide workers with exposures of more than 12 years (with a mean of $13^1/_3$ year) is still declining and was 0.0081 μg/ml for dieldrin only, with a top-level of 0.060 μg/ml. For all insecticides combined the mean concentration was 0.0230 μg/ml D.Eq. with a top-level of 0.120 μg/ml. For dieldrin this is 13 and 100 times the present blood level of the general population in the U.K. and the U.S.A. respectively. For all insecticides it is 38 and 200 times the D.Eq. of this general population respectively. From Table 20 we can extrapolate that 5 years ago the mean blood levels were nearly 3 times as high as these more recent values, and from other medical and technical information we know that during the earlier exposure years of this group, exposures must have been still higher.

We may therefore conclude from this part of the study, that our insecticide workers had patterns of health and rates of absenteeism due to disease and accidents, which were not different from operators working in other chemical plants.

The only changes in the parameters examined which might be attributed to insecticide exposure were slight increases of SGOT and SGPT with increasing D.Eq. — although all values remained within normal limits — and a statistically significant increase of serum alkaline phosphatase means — also within normal limits — in the D.Eq. groups 0.20-0.30 μg/ml, thus above our safety threshold level. In view of what was stated in chapters 2, 4 and 6 regarding these enzymes, this last phenomenon should be regarded as the first sign of adaptation of the liver cells to an insecticide exposure giving rise to insecticide concentrations in the blood above our safety threshold level, thus above blood levels of at least 333 times those of the general population in the U.K. and the U.S.A.

21.4   OCCUPATIONAL EXPOSURE TO ALDRIN AND DIELDRIN

The half-life of dieldrin in the blood − decreasing from a state of equilibrium in the body − was calculated to be 0.73 years in male workers. This is somewhat shorter than the 369 days, which Hunter *et al.* (1969) found in human volunteers.

The threshold level below which signs or symptoms of intoxication do not occur is in our experience at least 0.20 μg/ml in the blood for dieldrin. This is in agreement with data from the medical literature discussed in chapter 5 and it applies to all our insecticide workers, 800-900 men over an observation period of 15 years. This treshold level corresponds to at least 333 times the present blood level of dieldrin in the general population in the U.S.A. and the U.K. The data suggest that the no-effect-level in man corresponds to at least 0.20 μg/ml in the blood for dieldrin.

In the rat, dog, and monkey the no-effect-level for hepatic enzyme induction was found to be a daily intake of 50μg/kg. In man this would correspond to a daily intake of 3000-3500 μg, or 430-500 times the present daily intake of the general population in the U.S.A. and the U.K. (chapter 5.7), which is an intake which would result in blood levels exceeding our safety threshold level. At those high intakes liver enlargement was found in the rat and the dog. In man however, even in cases of intoxication, no hepatic enlargement was found on physical examination (14.2.1).

In the study of parameters of enzyme induction (chapter 20) it could be demonstrated that in man dieldrin levels in the blood of at least 175 times that of the general population did not affect the metabolism of pp'DDE as seen from pp'DDE levels in the blood. Dieldrin levels in the blood of at least 85 times that of the general population did not have an effect on steroid metabolism as measured in the ratio of 6-β-hydroxy-cortisol /17-hydroxy-corticosteroid excretion in the urine. This was also true for the man in this group, who had the highest dieldrin level of 0.110 μg/ml in the blood, which is more than 175 times that of the general population.

Thus, in blood levels up to at least 175 times those of the general population, or, in fact, at 0.105 μg/ml dieldrin in the blood, no signs were found of enzyme induction in our workers with the methods used. This does not mean that this is really the highest no-effect-level for these responses, but only that at these, highest presently studied exposures, these effects could not be demonstrated.

Thus, the total of these data shows that 0.105 μg/ml dieldrin in the blood − or 175 times the present blood level in the general population of the U.S.A. and the U.K. − is still below the human no-effect-level for dieldrin with the parameters used in this study. This no-effect-level might well be higher; in workers with higher blood levels up to 0.200 μg/ml dieldrin in the blood − or 333 times the blood level of the general population − on whom at that time no enzyme induction tests were performed, we found no effect in all other parameters examined.

21.5   OCCUPATIONAL EXPOSURE TO ENDRIN

In our opinion occupational exposure to endrin in our insecticide plant is of the same order as exposure to dieldrin. This, however, does not in normal circumstances, give rise to endrin levels in the blood above our detection level of 0.005 μg/ml. Only shortly after acute over-exposures endrin could be detected. Thus the non-persistence of endrin in the body is, in view of its pharmacokinetics, confirmed by our studies.

From our own data and those in the literature the toxic threshold level for endrin is suggested to be approximately 0.050-0.100 μg/ml, and the biological half-life of endrin in human blood would appear to be about 24 hours.

In the study of parameters of enzyme induction significantly lower pp'DDE levels were found in our endrin workers, and also in former endrin workers. Even 5 years or more after cessation of endrin work pp'DDE levels were still significantly lower than those of the general population.

6-β-hydroxycortisol excretion as expressed in the ratio of 6-β-hydroxycortisol/ 17-OH-corticosteroids in the urine was examined in 8 endrin workers. Significantly higher excretions were found than in non-exposed and in aldrin-dieldrin exposed workers. This is to be regarded as an expression of enhanced cortisol hydroxylation.

As endrin exposure cannot be quantified exactly, it is not possible to relate this enhanced metabolism of pp'DDE and cortisol to the degree of endrin exposure. This finding is of importance for workers occupationally exposed to endrin. For the general population this finding is irrelevant, since endrin is not normally found in food, air, and water.

These data show that occupational exposure in endrin manufacturing may cause enzyme induction in hydroxylating enzyme systems, although this response in no way affects the health of the endrin workers. From the available data it is not certain whether endrin or one of the intermediates in endrin manufacturing is the causitive agent. In other groups of workers this phenomenon was not demonstrated. The general population is not normally exposed to endrin.

21.6 OCCUPATIONAL EXPOSURE TO TELODRIN

Although Telodrin production ceased more than 4 years ago, this insecticide still had to be considered because of its long biological half-life.

The safety threshold level for Telodrin in the blood in our experience is 0.015 μg/ml.

The half-life of Telodrin in blood, at the state of equilibrium in the body, was calculated to be 2.77 years.

Signs and symptoms of Telodrin intoxication in some cases persisted for 6 months after the accident.

Notwithstanding the high acute toxicity of Telodrin and the high persistence of both the presence of the insecticide in the body, and its clinical effects after intoxication, we found no persistent adverse after-effects of exposure to this insecticide in our workers.

21.7 CONCLUSION

Apart from typical effects of intoxication after gross over-exposure, only a few adaptive responses of liver cells were found at insecticide levels in blood above our safety thresholds. Signs of enzyme induction were found in former and present endrin workers.

Workers with typical signs and symptoms of insecticide intoxication which occurred during the earlier production years fully recovered.

In human aldrin and dieldrin exposure we established with the methods used a no-effect blood level of at least 0.105 μg/ml, or at least 175 times the blood level found in the general population in the U.S.A. and the U.K. It is suggested that this no-effect-level is a conservative one. Since there were no workers with higher exposures it could not be demonstrated that in man the no-effect-level is still higher than 0.105 μg/ml.

Endrin workers with a daily exposure to endrin estimated to be of the same order as dieldrin exposure in dieldrin workers, but which cannot be quantified on account of the rapid metabolism and excretion of endrin, were shown to have an enhanced activity of hydroxylating hepatic enzyme systems, however, without any clinical manifestation. This effect was shown to be a persistent one, for which no explanation can be given at this time. The phenomenon of enzyme induction in present and former endrin workers in all its aspects, needs further study.

For the general population, these results mean that at the present level of exposure to aldrin and dieldrin in the U.S.A. and the U.K., and probably in most of Western Europe also, there is a safety margin of at least 175 times between this exposure and an industrial exposure at which neither general clinical effects or biochemical responses, nor signs of enzyme induction could be demonstrated by all methods used in this study, even after 10-15 years exposure.

Since the general population is not normally exposed to endrin, the enzyme induction found in endrin workers is not relevant for the general population.

Chapter 22

SUMMARY

22.1 SAMENVATTING

In de insecticidenfabriek van Shell Nederland Chemie N.V. te Pernis worden sinds eind 1954 de gechloreerde cyclische koolwaterstoffen aldrin, dieldrin en endrin geproduceerd. Gedurende een kortere periode tot 1965 werd ook ''TELODRIN''* gemaakt. Deze vier verbindingen zijn cyclodieen-derivaten, welke toepassing vinden als insecticiden. Sinds het begin hebben bijna 900 werknemers voor kortere of langere tijd in deze insecticidenfabriek gewerkt.

Het doel van dit onderzoek is een evaluatie van alle beschikbare medische gegevens van deze werknemers, ten einde na te gaan of het langdurig intensief contact met deze insecticiden de gezondheidstoestand van de werknemers op enigerlei wijze beïnvloed heeft. In tweede instantie is het dan mogelijk om uit deze gegevens van een groep industrieel blootgestelden conclusies te trekken ten aanzien van de veiligheid en de aanvaardbaarheid van de — uiteraard veel minder intensieve — blootstelling van de bevolking in het algemeen.

Als inleiding tot dit onderzoek wordt allereerst een kort overzicht gegeven van het nut zowel als van de gevaren voor mens en milieu van het gebruik van pesticiden.

Vervolgens worden een aantal toxicologische principes, voor zover van belang voor dit onderzoek, besproken. De parameters die in dit onderzoek gebruikt worden voor het aantonen van eventuele effecten in het lichaam worden toegelicht. Deze parameters zijn zó gekozen, dat niet alleen duidelijke stoornissen in de functie van organen worden ontdekt, maar dat ook de eerste adaptatieverschijnselen van de lichaamscellen aan de toxische stof, en de door deze stof opgewekte enzyminductie, indien aanwezig, kan worden geregistreerd in een phase, die nog niet als schadelijk voor het lichaam aangemerkt behoeft te worden.

Al deze parameters worden waar mogelijk gerelateerd aan de mate van blootstelling, die in het geval van aldrin, dieldrin en Telodrin gemeten kan worden aan de concentratie van dieldrin en Telodrin in het bloed. De bepaling van deze insecticidenconcentraties in het bloed van de werknemers met een geperfectioneerde gaschromatografische techniek is momenteel de beste methode voor biologische controle. Hierbij wordt namelijk zowel de

---

*''TELODRIN'' is een handelsmerk van Shell. Ter vereenvoudiging wordt in het vervolg Telodrin geschreven.

gemiddelde dagelijkse opname via alle opnamewegen, als de reeds bestaande "body burden" gemeten.

Een overzicht wordt gegeven van de chemische structuur en van enkele relevante physisch-chemische eigenschappen van deze insecticiden en van de daaruit geformuleerde handelsproducten.

Tevens wordt een literatuuroverzicht gegeven van het toxicologisch onderzoek bij mens en dier. Van de conclusies die hieruit getrokken kunnen worden zijn op deze plaats relevant:

— Aldrin wordt in het lichaam snel geëpoxideerd tot dieldrin. Met de dieldrinspiegel in het bloed worden dus de aldrin en de dieldrin blootstelling samen gemeten.

— Voor de mens is de relatie bekend, die bestaat tussen de gemiddelde dagelijkse opname van aldrin en dieldrin en de daardoor bereikte evenwichtsconcentratie in bloed en vetweefsel.

— Na het beëindigen van de blootstelling aan deze insecticiden dalen de bloedspiegels en de concentratie in vetweefsel weer exponentieel, zoals ze aanvankelijk gestegen zijn.

— De blootstelling van de bevolking in het algemeen aan aldrin en dieldrin in de Verenigde Staten, Engeland, en waarschijnlijk ook in West Europa, vertoont een dalende tendens en is volgens de laatste gegevens in de orde van grootte van 0.1 μg/kg/dag, of 7 μg/man/dag, overeenkomend met een evenwichtsconcentratie van dieldrin in bloed van 0.0006 μg/ml.

— Ondanks het feit, dat endrin een stereo-isomeer van dieldrin is, onderscheidt het zich van dit laatste insecticide door een grotere acute toxiciteit en een sneller metabolisme in, en eliminatie uit, het lichaam. Endrin is niet persistent in gewervelde dieren.

— De vier genoemde insecticiden geven bij een acute intoxicatie alle hetzelfde beeld, namelijk symptomen van overprikkeling van het centrale zenuwstelsel.

Er wordt een korte beschrijving gegeven van de insecticidenfabriek, speciaal wat betreft aspecten als veiligheidsvoorzieningen, hygienische voorzieningen en maatregelen ter voorkoming van besmetting in de fabriek en van het milieu. Ook worden de verschillende taken van de werknemers in deze fabriek in het kort belicht, evenals hun selectie en opleiding. Het medische routine onderzoek van deze werknemers wordt beschreven. Uit de beschrijving van de toegepaste medische selectie blijkt, dat de groep onderzochte werknemers van de insecticidenfabriek uit medisch oogpunt niet essentieel verschilt van elke andere groep industriële werknemers.

Dit alles vormt de inleiding en de achtergrond waartegen het epidemiologisch-toxicologisch onderzoek van de werknemers van de insecticidenfabriek reliëf krijgt. Behalve de evaluatie van alle beschikbare medische gegevens van deze groep werknemers worden speciale studies gewijd aan al die werknemers, die langer dan vier jaar in deze fabriek gewerkt hebben, en aan de 54 gevallen van insecticide-intoxicatie, die zich in de eerste productiejaren voorgedaan hebben.

Daarnaast worden 20 parameters met betrekking tot de algemene gezondheids-toestand en verschillende orgaanfuncties onderzocht, alsmede enkele parameters van enzyminductie. Tenslotte wordt het verzuim ten gevolge van ziekten en ongevallen aan een nader onderzoek onderworpen.

De belangrijkste resultaten hiervan kunnen als volgt samengevat worden:

1. Bij werknemers die beroepshalve met deze insecticiden in aanraking komen — soms

zelfs in toxische doses, en in tijdsduur oplopend tot 15 jaar — zijn met alle gebruikte parameters geen schadelijke effecten op de gezondheidstoestand gevonden.

2. Alle klachten en symptomen die optreden bij intoxicaties door deze insecticiden zijn volledig reversibel: in het geval van endrin binnen enkele dagen, bij aldrin en dieldrin binnen enkele weken, en in het geval van Telodrin binnen maximaal 6 maanden.

3. De grenswaarden van de insecticidenconcentraties in het bloed, waaronder nooit klachten of symptomen als gevolg van blootstelling aan deze insecticiden optreden, zijn op grond van onze gegevens:

0.20 μg/ml voor dieldrin, hetgeen overeenkomt met 333x de huidige dieldrinconcentratie in het bloed van de bevolking in het algemeen in de Verenigde Staten en Engeland en

0.015 μg/ml in het geval van Telodrin,

bij endrin ligt deze grenswaarde waarschijnlijk tussen 0.050-0.100 μg/ml.

4. De biologische halfwaardetijd van deze insecticiden, uitgaande van een evenwichtstoestand in het lichaam, wordt berekend op:

0.73 jaar voor dieldrin,

2.77 jaar voor Telodrin,

voor endrin kan ze worden geschat op ongeveer 24 uur.

5. De percentages en het patroon van het verzuim ten gevolge van ziekten en ongevallen van de werknemers in de insecticidenfabriek verschillen niet van die van werknemers in onze andere chemische fabrieken.

6. Bij een blootstelling aan aldrin en dieldrin overeenkomend met een evenwichtsconcentratie van 0.105 μg/ml dieldrin in het bloed — de hoogste concentratie waarbij dit kon worden nagegaan — worden met de gebruikte methoden geen tekenen van enzyminductie gevonden. Tot aan een evenwichtsconcentratie van 0.200 μg/ml dieldrin in het bloed worden geen afwijkingen gevonden in alle andere onderzochte parameters. Het "no-effect-level" voor aldrin en dieldrin is, wanneer rekening gehouden wordt met alle onderzochte parameters, tenminste 0.105 μg/ml dieldrin in het bloed. Het is echter zeer wel mogelijk dat het "no-effect-level" hoger ligt.

7. De blootstelling aan endrin kan tot nu toe niet gequantificeerd worden in een bloedspiegel, zoals dat gebeurt bij dieldrin. Bij de in de endrinproductie beroepshalve blootgestelde werknemers worden geen klinische afwijkingen gevonden, en ook geen afwijkingen in alle in het laboratorium onderzochte parameters. Bij deze groep worden echter wel duidelijke aanwijzingen gevonden voor het optreden van enzyminductie.

8. De huidige blootstelling van de bevolking in het algemeen aan aldrin en dieldrin in Engeland en de Verenigde Staten — en waarschijnlijk ook in het grootste deel van West Europa — resulteert in een dieldrinspiegel in het bloed die 175 x lager is dan de bij de mens vastgestelde "no-effect-level". Deze blootstelling mag dus als veilig beschouwd

worden, voor zover dat geëvalueerd kan worden met alle in dit onderzoek gebruikte parameters.

9. De bevolking in het algemeen is normaal niet blootgesteld aan endrin. Dus zijn de bij de werknemers van de endrinfabriek gevonden aanwijzigingen voor het optreden van enzyminductie niet van belang voor deze bevolking.

10. De bevolking in het algemeen is niet blootgesteld aan Telodrin. In verband met het feit, dat de Telodrinproductie al in 1965 werd gestaakt, zijn de resultaten van het onderzoek voor zover die betrekking hebben op Telodrin alleen van wetenschappelijk en historisch belang.

22.2 SUMMARY*

A brief introduction to the various health aspects of the use of pesticides is given, followed by a discussion of the relevant toxicology and of some general principles of toxicology. In this study various parameters were used in an attempt to demonstrate whether or not responses to the measured insecticide exposure occurred. The method of choice of monitoring people exposed to organochlorine insecticides proved to be the determination of insecticide levels in the blood by gas liquid chromatography with an electron capture detector.

Some general physico-chemical information on the cyclodiene insecticides aldrin, dieldrin, endrin and "TELODRIN"** is given and animal and human toxicological data from the literature are reviewed and discussed.

A brief description of the insecticide plant of Shell Nederland Chemie N.V. is given, particularly from a point of view of personal safety and hygiene and precautions taken against internal and environmental contamination. The methods of routine medical examination of insecticide workers are described.

The above information may be regarded as an introduction to the epidemiological-toxicological study of the nearly 900 workers who have been occupationally exposed in our insecticide plant for shorter or longer periods during the last 15 years. Special studies were made of those exposed for more than 4 years and of all workers who suffered insecticide intoxications. Parameters of general health and of enzyme induction and of absenteeism due to disease and accident were also subject of special studies.
The results of all these studies are summarized as follows:

1. Occupational exposure to these insecticides for periods up to 15 years, at times even at toxic doses, did not have any persistent adverse effect on the health of these workers as far as could be demonstrated in all parameters used.

2. In cases of intoxication all signs and symptoms were fully reversible: in the case of endrin within a few days, with aldrin and dieldrin within weeks, and in cases of Telodrin intoxication within 6 months.

---

* Note: at the end of each chapter more extensive summaries have been given.
**"TELODRIN" is a Shell trade mark. For ease of reading :"TELODRIN" shall be written Telodrin.

3. The safety threshold levels of insecticide concentrations in the blood below which no signs or symptoms of intoxication ever occurred are, in our experience:

0.20 μg/ml for dieldrin, which is 333 times the present dieldrin level in the blood of the general population in the U.K. and the U.S.A.,

0.015 μg/ml for Telodrin, and

for endrin it is believed to be 0.050-0.100 μg/ml.

4. The biological half-life of these insecticides, decreasing from a state of equilibrium in the human body, is calculated to be:

0.73 years for dieldrin,

2.77 years for Telodrin and

for endrin is estimated to be approximately 24 hours.

5. The pattern and rates of absenteeism due to disease and accident of our insecticide workers are similar to that of operators of the same average age in other chemical plants.

6. Aldrin and dieldrin exposures up to an equilibrium blood level of 0.105 μg/ml were shown to cause no signs of enzyme induction. Up to a blood level of 0.200 μg/ml there occurred no clinical effects and no changes in the parameters examined. A no-effect-level in man for aldrin and dieldrin with all parameters used was established at a blood level of at least 0.105 μg/ml dieldrin and might probably be higher.

7. Endrin exposure could not be quantified with blood levels in the same way as could be done with dieldrin. Occupational exposure to endrin caused no adverse clinical effects, or changes in the general health parameters examined. However, enzyme induction was clearly demonstrated in endrin manufacturing workers.

8. The present exposure of the general population to aldrin and dieldrin in the U.K. and the U.S.A. – and probably also in most of Western Europe – results in a blood level which is 175 times lower than the established no-effect-level in man. This exposure may thus be regarded as safe, as far as could be demonstrated with the parameters used in this study.

9. The general population is not normally exposed to endrin. Thus the signs of enzyme induction found in endrin manufacturers are not relevant for the general population.

10. The general population is not exposed to Telodrin. Since Telodrin manufacturing ceased in 1965, the results mentioned concerning Telodrin are of scientific and historical interest only.

22.3 RESUME

Une brève introduction est donnée aux différents aspects sanitaires de l'usage de pesticides, suivie d'un exposé sur les propriétés toxicologiques de ces produits et sur quelques principes généreaux de la toxicologie.

Dans cette étude plusieurs paramètres ont été utilisés dans une tentative de montrer si à certains degrés mesurés d'exposition aux insecticides des réactions sont provoquées ou non. Il s'est trouvé que la méthode à préférer pour surveiller des personnes exposées à des insecticides organochlorurés, est la détermination des teneurs en insecticides du sang par chromatographie gaz-liquide à l'aide d'un détecteur de capture électronique.

La thèse contient quelques données physico-chimiques généraux sur l'aldrin, le dieldrin, l'endrin et le "TELODRIN"*, insecticides à base de cyclodiène, ainsi qu'un aperçu et une discussion de données toxicologiques concernant l'animal et l'homme.

Une description sommaire est donnée de l'usine d'insecticides de Shell Nederland Chemie N.V., notamment du point de vue de la sécurité et de l'hygiène personnelles et des mesures préventives prises contre la contamination à l'intérieur de l'usine et celle de l'environnement. On décrit aussi les méthodes des examens médicaux de routine auxquels sont soumis les ouvriers travaillant dans des usines d'insecticides.

Les informations précédentes peuvent être considérées comme une introduction à l'étude épidémio-toxicologique concernant les près de 900 ouvriers qui, travaillant dans l'usine d'insecticides, ont été exposés pendant des périodes plus ou moins longues au cours des 15 dernières années. Des examens spéciaux ont été faits chez ceux exposés pendant plus de 4 années et chez tous les ouvriers souffrant d'intoxications par des insecticides.

Des paramètres de l'état général de santé, de l'induction enzymatique, de l'absence pour cause de maladies et d'accidents, ont également fait l'objet d'études spéciales.

Les résultats de toutes ces études peuvent se résumer comme suit:

1. Pour autant qu'on a pu le montrer dans tous les paramètres utilisés, l'exposition à ces insecticides par suite du travail, pendant des périodes allant jusqu'à 15 années, parfois même à des doses toxiques, n'a eu aucun effet contraire persistant sur la santé de ces ouvriers.

2. Dans les cas d'intoxication, tous les signes et symptômes ont été complètement réversibles, c.-à-d. en quelques jours dans le cas de l'endrin, en quelques semaines dans celui de l'aldrin et du dieldrin, et en six mois quand il s'agissait d'une intoxication par du Telodrin.

3. Les seuils des teneurs en insecticide du sang, au-dessous desquels aucun signe ou symptôme d'intoxication ne s'est jamais manifesté, sont, d'après notre expérience:

de 0.20 μg/ml pour le dieldrin, soit 333 fois la teneur actuelle en dieldrin du sang de la population générale du Royaume-Uni et des Etats-Unis,

de 0.015 μg/ml pour le Telodrin,

alors que pour l'endrin on suppose qu'elle se situe entre 0.050 et 0.100 μg/ml.

4. La durée de la demie-vie de ces insecticides, qui décroit dès que l'état d'équilibre dans le corps humain n'existe plus, est estimée à:

---

*"TELODRIN" est une marque déposée Shell. Pour faciliter la lecture, on écrira Telodrin au lieu de "TELODRIN".

220

0.73 années pour le dieldrin,
2.77 années pour le Telodrin,
alors que dans le cas de l'endrin on suppose qu'elle est de quelque 24 heures.

5. La structure et la fréquence de l'absentéisme dû à des maladies et à des accidents des ouvriers de l'usine d'insecticides, sont comparables à celles enregistrées pour des travailleurs d'un même âge moyen dans d'autres usines chimiques.

6. On a trouvé que l'exposition à l'aldrin et au dieldrin jusqu'à correspondance d'une teneur de 0.105 µg/ml du sang en dieldrin à l'état d'équilibre, ne produit pas des signes d'induction enzymatique. Jusqu'à une teneur de 0.200 µg/ml du sang, il ne s'est pas manifesté d'effects cliniques ni de changement dans les paramètres. Une teneur du sang en dieldrin n'ayant pas d'effet chez l'homme, d'après tous les paramètres utilisés, a été établie à 0.105 µg/ml au minimum, et elle est très susceptible d'être plus élevée.

7. L'exposition à l'endrin n'a pas pu être quantifiée par des teneurs en insecticide du sang de la même manière qu'on a pu le faire dans le cas du dieldrin. L'exposition à l'endrin par suite du travail n'a pas eu d'effects cliniques, ni a-t-elle provoqué de changements dans les paramètres considérés de l'état général de santé. En revanche, l'induction enzymatique a été clairement montrée chez des ouvriers engagés dans la fabrication d'endrin.

8. L'exposition actuelle à l'aldrin et au dieldrin de la population générale du Royaume-Uni et des Etats-Unis — et probablement de celle de la plus grande partie de l'Europe Occidentale — produit une teneur en insecticide du sang qui est 175 fois plus petite que celle établie comme n'ayant pas d'effet chez l'homme. Ce degré d'exposition peut donc être qualifié d'inoffensif, pour autant qu'on a pu prouver ceci à l'aide des paramètres utilisés dans cette étude.

9. La population générale n'étant normalement pas exposée à l'endrin, les signes d'induction enzymatique trouvés chez des ouvriers fabriquant de l'endrin ne s'appliquent pas à elle.

10. La population générale n'est pas exposée au Telodrin. Etant donné qu'on a arrêté en 1965 la fabrication de Telodrin, les résultats mentionnés concernant ce produit revêtent un intérêt purement scientifique et historique.

22.4 ZUSAMMENFASSUNG

Es wird eine kurze Einführung in die verschiedenen Gesundheitsaspekte bei der Verwendung von Schädlingsbekämpfungsmitteln gegeben mit anschliessend einer Besprechung der entsprechenden Toxikologie und einiger allgemeinen Grundsätze der Toxikologie. Bei dieser Untersuchung wurden in einem Versuch, herauszufinden ob bei gewissen gemessenen Einwirkungsgraden gegebenenfalls Reaktionen auftraten, verschiedene Parameter verwendet. Als die beste Methode, Betriebspersonal, das der Einwirkung von organischen chlorhaltigen Insektiziden ausgesetzt ist, zu überwachen, erwies sich die Bestimmung von im Blut befindlichen Insektizidmengen durch Gaschromatographie mit einem "Electron-capture"-Detektor.

Es wird eine allgemeine Charakteristik der Cyclodien-Insektizide Aldrin, Dieldrin, Endrin und "TELODRIN"* gegeben; ausserdem wird eine Übersicht der aus der Literatur bekannten toxikologischen Daten über Mensch und Tier gegeben und besprochen.

Die Anlagen der Shell Nederland Chemie N.V. zur Herstellung dieser Insektiziden werden kurz beschrieben, besonders vom Gesichtspunkt der persönlichen Sicherheit und Hygiene und der Vorsichtsmassregeln, die gegen die Betriebs- und Umgebungsveruntreinigung getroffen wurden. Es werden die Methoden beschrieben, die bei der routinemässigen ärztlichen Untersuchung von mit Insektiziden in Berührung kommenden Betriebspersonal angewendet werden.

Obige Angaben können als eine Einführung in die epidemiologische toxikologische Untersuchung der fast 900 Mann Betriebspersonal betrachtet werden, die während der letzten 15 Jahre kürzere oder längere Zeit in obengenannten Anlagen beruflich der Einwirkung von Insektiziden ausgesetzt waren. Besondere Studien wurden durchgeführt über diejenigen, die dieser Einwirkung über 4 Jahre ausgesetzt waren und über alle jene Arbeiter, die an Insektizid-Vergiftungen litten. Parameter der allgemeinen Gesundheit sowie Enzyminduktion und Arbeitsversäumnis durch Krankheiten und Unfälle waren ebenfalls Gegenstand spezieller Studien.

Die Ergebnisse all dieser Studien werden wie folgt zusammengefasst:

1. Sofern mit den verwendeten Parametern nachgewiesen werden konnte. hatte die Dauereinwirkung der genannten Insektizide, gelegentlich sogar in giftigen Mengen, während 15 Jahren, auf in den Herstellungsanlagen beschäftigtes Personal, keinen einzigen dauernden nachteiligen Einfluss auf die Gesundheit dieses Personals.

2. In Vergiftungsfällen waren alle Anzeichen und Symptome völlig reversibel. Bei Endrin trat völlige Rückbildung innerhalb von wenigen Tagen ein, bei Aldrin und Dieldrin innerhalb einigen Wochen, und bei Telodrinvergiftung nach höchstens 6 Monaten.

3. Die Schwellenwerte für Insektizidkonzentrationen im Blut, unter welchen keinerlei Vergiftungsanzeichen oder -symptome auftraten sind nach unserer Erfahrung:
0.20 μg/ml für Dieldrin, was das 333-fache der heutigen Dieldrin-Menge im Blut der Gesamtbevölkerung in Grossbritannien und den Vereinigten Staaten darstellt,
0.015 μg/ml für Telodrin,
für Endrin wird 0.050-0.100 μg/ml angenommen.

4. Als biologische Halbwertzeit dieser Insektizide, ausgehend von einem Gleichgewichtszustand im menschlichen Körper, werden folgende Werte erhalten:
für Dieldrin 0.73 Jahre,
für Telodrin 2.77 Jahre,
für Endrin werden ungefähr 24 Stunden geschätzt.

---

*"TELODRIN" ist ein Warenzeichen der Shell. Der Einfachheit halber ist diese spezielle Kennzeichnung im weiteren Text weggelassen.

5. Das Bild und die Frequenz der Arbeitsversäumnis als Folge von Krankheiten und Unfällen gleichen bei unserem Betriebspersonal, das mit diesen Insektiziden in Berührung kommt, völlig denjenigen eines vergleichbaren Durchschnitts in anderen chemischen Anlagen.

6. Gefunden wurde, dass die Einwirkung von Aldrin und Dieldrin bis zu einem Gleichgewichtsniveau im Blut von 0.105 μg/ml keine Anzeichen einer Enzyminduktion hervorrief. Bis zu einem Gehalt im Blut von 0.200 μg/ml ergaben sich keine klinische Auswirkungen und keine Änderungen in den hierbei untersuchten Parametern. Mit allen verwendeten Parametern wurde für den Menschen bei Aldrin und Dieldrin ein "Schwellenwert ohne Effect" von mindestens 0.105 μg/ml Dieldrin im Blut gefunden, der jedoch wahrscheinlich noch höher sein dürfte.

7. Die quantitative Einwirkung von Endrin konnte im Gegensatz zu derjenigen von Dieldrin nicht in gleicher Weise auf Basis des Blutgehaltes bestimmt werden. Die berufliche Daueraussetzung an die Einwirkung von Endrin verursachte keinerlei ungünstigen klinischen Auswirkungen oder Änderungen in den untersuchten allgemeinen Gesundheitsparametern. Dagegen konnte deutlich Enzyminduktion nachgewiesen werden.

8. Die heutige Aussetzung der Gesamtbevölkerung von Grossbritannien und den Vereinigten Staaten — und wahrscheinlich auch des grössten Teils von Westeuropa — an Aldrin und Dieldrin, ergibt einen Gehalt im Blut, der 175 mal niedriger ist als der festgestellte "no-effect"-Gehalt im Menschen. Soweit mit den bei der vorliegenden Untersuchung verwendeten Parametern festgestellt werden konnte, kann diese Einwirkung daher als gefahrlos betrachtet werden.

9. Die Bevölkerung ist normalerweise der Einwirkung von Endrin nicht ausgesetzt. Die bei den mit der Endrinverarbeitung beschäftigten Arbeitern gefundenen Anzeichen von Enzyminduktion sind daher für die Gesamtbevölkerung ohne Bedeutung.

10. Ebensowenig ist und war die Bevölkerung normalerweise der Einwirkung von Telodrin ausgesetzt. Da die Verarbeitung von Telodrin bereits 1965 eingestellt wurde, sind die betreffenden Resultate nur noch von wissenschaftlicher und historischer Bedeutung.

22.5 RESUMEN

Tras una breve introducción, dedicada a los diversos aspectos sanitarios relacionados con el uso de pesticidas, sigue una exposición de las propiedades toxicológicas de dichos productos y de algunos principios generales de la toxicología. En este estudio se han empleado varios parámetros en una tentativa de comprobar si ciertos grados de exposición a insecticidas, determinados mediante mediciones, provocaban o no determinadas reacciones. La determinación de los niveles de insecticida contenidos en la sangre, mediante cromatografía de fase gas — líquido con ayuda de un detector de captura de electrones, ha resultado ser el método preferible para vigilar a personas expuestas a insecticidas organoclorados.

La tesis contiene alguna información físicoquímica general sobre los insecticidas ciclodiénicos aldrín, dieldrín, endrín y "TELODRIN"*. Además se recogen y se comentan en ella algunos datos toxicológicos obtenidos en animales y en el hombre provenientes de la literatura del ramo.

La planta de insecticidas de Shell Nederland Chemie N.V. es objeta de una breve descripción en la que se resalta especialmente la seguridad e higiene personales y las precauciones tomadas para contrarrestar la contaminación al interior de la planta y la del medio ambiente. Se describen los métodos de reconocimiento médico periódico de los obreros de la fábrica de insecticidas.

La información anterior puede considerarse como una introducción al estudio epidemiotoxicológico de los casi 900 operarios que trabajan en dicha planta de insecticidas y que, en razón de su oficio, han estado expuestos durante periódos más o menos largos en el curso de los últimos 15 años. Se han hecho estudios especiales de los obreros que han estado expuestos durante más de 4 años y de todos los que sufrieron intoxicaciones originadas por insecticidas. También han sido objeto de estudios especiales algunos parámetros de la salud general, de la inducción enzimática y las faltas al trabajo debidas a enfermedades y accidentes.

Los resultados de todos estos estudios han sido resumidos en la forma siguiente:

1. En tanto se ha podido comprobar en todos los parámetros empleados, la exposición profesional a los referidos insecticidas durante periódos de hasta 15 años, a veces incluso a dosis tóxicas, no tuvo ningún efecto desfavorable persistente en la salud de los obreros en cuestión.

2. En los casos de intoxicación todos los indicios y síntomas han resultado ser completamente reversibles; en el caso de endrín en pocos días, en aldrín y dieldrín en algunas semanas y, en casos de intoxicación por Telodrín, en 6 meses.

3. Los niveles límites (umbrales) de concentraciones de insecticida en la sangre, debajo de los cuales no se produjeron jamás indicios o síntomas de intoxicación, se sitúan, de acuerdo con nuestra experiencia, en:

0.20 µg/ml para dieldrín, lo que representa 333 veces el nivel actual de dieldrín en la sangre de la población general del Reino Unido y de los Estados Unidos,

0.015 µg/ml para Telodrín,

mientras que para endrin se supone que se sitúa entre 0.050 y 0.100 µg/ml.

4. La vida media biológica de estos insecticidas, que va decreciendo en cuanto deja de existir el estado de equilibrio en el cuerpo humano, se calcula en

0.73 años para dieldrín

2.77 años para Telodrín

y para endrín se supone que es de 24 horas aproximadamente.

---

*"TELODRIN" es una marca registrada de Shell. Para mayor facilidad de lectura, "TELODRIN" escribirá Telodrin.

5. La imagen global y la frequencia de las faltas al trabajo debidas a enfermedades y accidentes de los operarios en la planta de insecticidas son similares a las registradas en otras fábricas de productos químicos en obreros de edad comparable.

6. Ha quedado demostrado que la exposición a aldrín y dieldrín hasta un nivel de equilibrio en la sangre de 0.105 μg/ml dieldrin no causa señales de inducción enzimática. Hasta un nivel en la sangre de 0.200 μg/ml no se manifestaron efectos patólogicos, ni ningún cambio en los parámetros correspondientes que se han examinado. Todos los parámetros empleados han permitido establecer un "nivel-sin-efecto" en el hombre para aldrín y dieldrín de por lo menos 0.105 μg/ml dieldrín de sangre, nivel que probablemente puede ser más elevado.

7. No se ha podido cuantificar la exposición a endrín a base de los niveles en la sangre de la misma manera que ello sí fue posible con dieldrín. La exposición profesional a endrín no causó efectos patológicos, ni cambios en los parámetros de la salud general examinados. Sin embargo, la inducción enzimática ha quedado claramente demostrada en operarios ocupados en la fabricación de endrín.

8. La exposición actual de la población general a aldrín y dieldrín en el Reino Unido y en los Estados Unidos, y probablemente también en la mayor parte de la Europa Occidental, conduce a un nivel en la sangre que es 175 veces inferior al "nivel-sin-efecto" establecido en el hombre. Esta exposición puede considerarse por lo tanto como inofensiva, en la medida en que los parámetros usados en este estudio han permitido comprobarlo.

9. Como la población general no está normalmente expuesta a endrín, los indicios de inducción enzimática encontrados en operarios ocupados en la fabricación de este producto no son significativos para la población general.

10. La población general no está expuesta a Telodrín. Como en 1965 se dio fin a la fabricación de Telodrín, los resultados mencionados respecto a este producto sólo revisten interés científico e histórico.

Chapter 23

REFERENCES

Aarts, E.M., Evidence for the function of D-glucaric acid as an indicator for drug induced enhanced metabolism through the glucuronic pathway in man (1965) *Biochem. Pharmacol.* **14**, 359-363.

Abbott, D.C., Harrison, R.B., Tatton, J.O'G. and Thomson, J. Organochlorine pesticides in the atmospheric environment (1965) *Nature* **208**, 1317-1318.

Abbott, D.C., Harrison, R.B., Tatton, J.O'G. and Thomson, J. Organochlorine pesticides in the atmospheric environment (1966) *Nature* **211**, 259-261.

Abbott, D.C., Goulding, R. and Tatton, J.O'G. Organochlorine pesticide residues in human fat in Great Britain (1968) *Brit. Med. J.* **3**, 146-149.

Ariëns, E.J. Verwachte en onverwachte reacties bij combinatie van geneesmiddelen (1969-a) *Ned. T. Geneesk.* **113**, 344-352.

Ariëns, E.J. Industriële toxicologie als onderdeel van de algemene toxicologie (1969-b). Paper read at the "Bedrijfsgeneeskundige Studiedag" 21 february 1969, Nijmegen.

Bäckström, J., Hanson, E. and Ullberg, S. Distribution of C^{14} DDT and C^{14} dieldrin in pregnant mice, determined by whole-body autoradiography (1965) *Tox. Appl. Pharmacol.* **7**, 90-96.

Baetjer, A.M. Changes-Stress or benefit? (1965) *Am. Ind. Hyg. Ass. J.* **25**, 207-212.

Bann, J.M., DeCino, T.J., Earle, N.W. and Sun, Y.P. The fate of aldrin and dieldrin in the animal body (1956) *J. Agric. Fd. Chem.* **4**. 937-941.

Barnes, J.M. and Heath, D.F. Some toxic effects of dieldrin in rats (1964) *Brit. J. Ind. Med.* **21**, 280-282.

Barnes, J.M. Carcinogenic hazards from pesticide residues (1966) *Residue Reviews* **13**, 69-82.

Barnes, J.M. The significance of pesticide residues in food (1967) *Span* **10**, 7-8.

Barnes, J.M. Food and health-The safe use of pesticides (1967) *Brit. Food. J.* 71-75.

Bass, S.W. and Triolo, A.J. Effect of DDT on parathion toxicity in mice (abstract) (1968) *Tox. Appl. Pharmacol* **12**, 289.

Bell, A. Aldrin poisoning-a case report (1960) *Med. J. Austr.* **2**, 698-700.

Benson, W.R. The chemistry of pesticides (1969) *Ann. N. York Acad. Sci.* **160**, Art. 1, 7-29.

Besemer, A.F.H. Bestrijdingsmiddelen, de balans van vóór- en nadelen (1965) Verslag van de 4e Drentse provinciale gezondsheidsdag: *De mens in zijn levensmilieu bedreigd?* Assen, Prov. Raad v.d. Volksgezondheid.

Bessey, O.A., Lowry, O.H. and Brock, M.J. A method for the rapid determination of alkaline phosphatase with five milliliters of serum (1946) *J. Biol. Chem.* **164**, *321-329*.

Bick, M. Chlorinated hydrocarbon residues in human body fat (1967) *Med. J. Austr.* 1127-1130.

Boorder, Tj. De and Ensberg, I.F.G. Carcinogene en haematotoxische effecten van insecticiden uit de chloorkoolwaterstofreeks (1969) *T. Soc. Geneesk.* **47**, 56-58.

Braund, D.G., Brown, L.D., Huber, J.T., Leeling, N.C. and Zabik, H.J. Placental transfer of dieldrin in dairy heifers contaminated during three stages of gestation (1968) *Journ. Dairy Science* **51**, 116-119.

Breidenbach, A.W. Pesticide residues in air and water (1965) *Arch. Environ. Health* **10**, 827-830.

Brewerton, H.V. and McGrath, H.J.W. Insecticides in human fat in New Zealand (1967) *N.Z. Jl. Sci.* **10**, 486-492.

Brown, J.R. Organochlorine pesticide residues in human depot fat (1967) *Can. Med. Ass. J.* **97**, *367-373*.

Brown, V.K.H., Chambers, P.L., Hunter, C.G. and Stevenson, D.E. The toxicity of Telodrin for vertebrates (1962) *Tunstall Laboratory Report* R (T)-2-62.

Brown, V.H.K., Hunter, C.G. and Richardson, A. A blood test diagnostic of exposure to aldrin and dieldrin (1964) *Br. J. Ind. Med.* **21**, 283-286.

Bruin, A. De and Zielhuis, R.L. Vroege diagnostiek in de bedrijfsgeneeskunde met behulp van biochemische methoden.

I: *T. Soc. Geneesk.* **45**, (1967) 128-134,

II: *T. Soc. Geneesk.* **45**, (1967) 163-170,

III: *T. Soc. Geneesk.* **45**, (1967) 196-198,

IV: *T. Soc. Geneesk.* **45**, (1967) 286-293.

Burns, J.J., Conney, A.H. and Koster, R. Stimulatory effect of chronic drugs administration on drug metabolising enzymes in liver microsomes (1963) *Annals N.Y. Acad. Sci.* **104**, 881-893.

226

Burstein, S. and Klaiber, E.L. Phenobarbital induced increase in 6-β-hydroxycortisol excretion: clue to its significance in human urine (1965) *J. Clin. Endocrinol.* **25**, 293-296.

Cabaud, P.G. and Wroblewski, F. Colorimetric measurement of lactic dehydrogenase activity of body fluids (1958) *Am. J. Clin. Pathol.* **30**, 234-236.

Campbell, J.E. Pesticide residues in human diet (1964) Paper read at the Meeting of American Chem. Society, Chicago, September 1964.

Campbell, J.E., Richardson, L.A. and Schafer, M.L. Insecticide residues in the human diet (1965) *Arch. Environ. Health* **10**, 831-836.

Carman, G.E. and DeBach, P. A rational approach to insect control (1967) *Span* **10**, 16-18.

Carson, R. *Silent Spring* – Houghton – Mifflin Co., Boston, Mass. (1962).

Casarett, L.J., Fryer, G.C., Yauger Jr. W.L. and Klemmer, H.W. Organochlorine pesticide residues in human tissue-Hawaii (1968) *Arch. Environ. Health* **17**, 306-331.

Cassidy, W., Fisher, A.J., Peden, J.D. and Parry-Jones, A. Organochlorine pesticide residues in human fats from Somerset (1967) Report in *Monthly Bull. Minist, Health* (London) **26**, 2-6.

Catz, C and Yaffe, S.J. Pharmacological modification of bilirubin conjugation in the newborn (1962) *Am. J. Dis. Child.* **104**, 516-517.

Chambers, P.L. The physiological and pharmacological effects of Telodrin (1962) *Tunstall Laboratory Report* R (T) 1-62.

Cleveland, F.P. A summary of work on aldrin and dieldrin toxicity at the Kettering Laboratory (1966) *Arch. Environ. Health* **13**, 195-198.

Coble, Y., Hildebrandt, P., Davis, J., Raasch, F. and Curley, A. Acute endrin poisoning (1967) *JAMA* **202**, 153-157.

Cole, J.F., Klevay, L.M., and Zavon, M.R. Endrin and dieldrin: a comparison of hepatic excretion rates in the rat (1968) Paper read at the 7th annual meeting of the Society of Toxicology, Washington D.C., March 4-6, 1968. Abstract in *Tox. Appl. Pharmacol.* **12**, 298.

Conney, A.H. Pharmacological implications of microsomal enzyme induction (1967) *Pharmacol. Rev.* **19**, 317-366.

Conney, A.H. Drug metabolism and therapeutics (1969) *N. Engl. J. Med.* **280**, 653-660.

Coulston, F. (1969) Paper read at Symposium on persistent pesticides-Albany Medical College, Albany, N.Y., U.S.A. 4 November 1969.

Cueto Jr., C. and Hayes Jr., W.J. The detection of dieldrin metabolites in human urine (1962) *J. Agr. Food Chem.* **10**, 366-369.

Cueto Jr., C. and Biros, F. J. Chlorinated insecticides and related materials in human urine (1967) *Tox. Appl. Pharmacol.* **10**, 261-269.

Cueto Jr., C., Hayes Jr., W.J. Effect of repeated administration of phenobarbital on the metabolism of dieldrin (1967) *Ind. Med. Surg.* **36**, 546-551.

Cummings, J.G. Total diet study-pesticide residues in total diet samples (1965) *Journ. A.O.A.C.* **48**, 1177-1180.

Curley, A. and Kimbrough, R. Chlorinated hydrocarbon insecticides in plasma and milk of pregnant and lactating women (1969) *Arch. Environ. Health* **18**, 156-164.

Dale, W.E. and Quinby. G.E. Chlorinated insecticides in the body fat of people in the United States (1963) *Science* **142**, 593-595.

Dale, W.E., Copeland, M.F. and Hayes Jr., W.J. Chlorinated insecticides in the body fat of people in India (1965) *Bull. Wld. Hlth. Org.* **33**, 471-477.

Dale, W.E., Curley, A. and Cueto Jr., C. Hexane extractable chlorinated insecticides in human blood (1966) *Life Sci.* **5**, 47-54.

Dale, W.E., Curley, A. and Hayes Jr., W.J. Determination of chlorinated insecticides in human blood (1967) *Ind. Med. Surg.* **36**, 275-280.

Damico, J.N., Chen, J.Y.T., Costello, C.E. and Haenni, E.O. Structure of Klein's metabolites of aldrin and dieldrin (1968) *Journ. A.O.A.C.* **51**, 48-55.

Datta, P.R., Laug, E.P., Watts, J.O., Klein, A.K. and Nelson, M.J. Metabolites in urine of rats on diets containing aldrin or dieldrin (1965) *Nature* **208**, 289-290.

Davies, G.M. and Lewis, I. Outbreak of food poisoning from bread made of chemically contaminated flour (1956) *Brit. Med. J.* **2**, 393-398.

Davies, J.E., Edmundson, W.F., Schneider, N.J. and Cassady, J.C. Pesticides in people (1968) *Pest. Monitoring J.* **2**, 80-85.

Davies, J. E., Edmundson, W. F., Carter, C. H. and Barquet, A. Effect of anticonvulsant drugs on Dicophane (DDT) residues in man (1969-a) *Lancet* **2**, 7-9.

Davies, J. E., Edmundson, W. F., Maceo, A., Barquet, A. and Cassady, J. An epidemiologic application of the study of DDE levels in whole blood (1969-b) *Amer. J. Publ. Hlth.* **59**, 435-441.

Davis, K. J. and Fitzhugh, O. G. Tumorigenic potential of aldrin and dieldrin for mice (1962) *Tox. Appl. Pharmacol.* 4, 187-189.

Decker, G. C. The significance of pesticide residues in the environment (1967) *Span* 10, 5-7.

Decker, G. C. Environmental aspects of using residual insecticides with special reference to aldrin and dieldrin (1968) Paper read at the symposium on the science and technology of residual insecticides in food production with special reference to aldrin and dieldrin-Gaithersburg, Md., U.S.A., June 1968.

Deichmann, W. B., Keplinger, M., Sala, F. and Glass, E. Synergism among oral carcinogens IV: the simultaneous feeding of four tumorigens to rats (1967) *Tox. Appl. Pharmacol.* 11, 88-103.

Deichmann, W. B., Dressler, I., Keplinger, M. and MacDonald, W. E. Retention of dieldrin in blood, liver and fat of rats fed dieldrin for six months (1968) *Ind. Med. Surg.* 37, 837-839.

Deichmann, W. B., Radomski, J. L., and Rey, A. Retention of pesticides in human adipose tissue-preliminary report (1968) *Ind. Med. Surg.* 37, 218-219.

Deichmann, W. B., Keplinger, M. L., Dressler, I. and Sala, F. Retention of dieldrin and DDT in the tissues of dogs fed aldrin and DDT individually and as a mixture (1969) *Toxicol. Appl. Pharmacol.* 14, 205-213.

Del Vecchio, V. and Leoni, V. La ricerca ed il dosaggio degli insetticidi clorurati in materiale biologico (1967) *Nuovi An.. Microbiol.* 28, 107-128.

Dijk, M. C. Van Chlorinated insecticides and the serum alkaline phosphatase (1968) Proc. 6th Shell Ind. Doct. Meeting, 28-30 May 1968 Amsterdam.

Dommelen, C. K. V. Van-De betekenis van het eiwitspectrum in de diagnostiek (1961) *Ned. T. Geneesk.* 105, 282-285.

Donninger, C. and Wright, A. S. Livercell changes, induced by the oral administration of dieldrin and phenobarbitone to rats (1967) *Tunstall Laboratory Report*, R (T)-10-67.

Duggan, R. E., Barry, H. C. and Johnson, L. Y. Pesticide residues in total diet samples (1966) *Science* 151, 101-104.

Duggan, R. E. and Dawson, K. Pesticides, a report on residues in food (1967-a) *FDA Papers* 1, 2-10.

Duggan, R. E. and Wetherwax, J. R. Dietary intake of pesticide residues (1967-b) *Science* 157, 1007-1010.

Durham, W. F. The interactions of pesticides with other factors (1967) *Residue Reviews* 18, 21-103.

Edmundson, W. F., Davies, J. E. and Hull, W. Dieldrin storage levels in necropsy adipose tissue from South Florida population (1968) *Pesticide Monit. J.* 2, 86-89.

Edmundson, W. F., Davies, J. E., Nachman, G. A. and Roeth, R. L.-P, p'-DDT and p, p'-DDE in blood samples of occupationally exposed workers (1969) *Publ. Hlth. Rep.* 84, 53-58.

Egan, H., Goulding, R., Roburn, J. and Tatton, J. O'G. Organochlorine pesticide residues in human fat and human milk (1965) *Brit. Med. J.* 2, 66-69.

Ely, R. E., Moore, L. A., Carter, R. H. and App. B. E. Excretion of endrin in the milk of cows fed endrin-sprayed alfalfa and technical endrin (1957) *J. Econ. Entomol.* 50, 348-349.

FAO/WHO Evaluation of the toxicity of pesticide residues in food (1965) *FAO Meeting Report* No. PL/1965/10/1 WHO/Food Add/27.65.

FAO/WHO Evaluation of some pesticide residues in food (1967) *FAO*, PL: CP/15, WHO/Food Add/67.32.

FAO/WHO *WHO Technical Report Series* No. 348 (1967) Geneva.

FAO/WHO Pesticide residues in food (1967) *Wld. Hlth. Org. Techn. Rep. Ser.* (1967) No. 370.

FAO/WHO Pesticide residues (1968), *Wld. Hlth. Org. Techn. Rep. Ser.* (1968) No. 391.

Flipse, L. P. and Schuddeboom, L. J. Overheidstoezicht op bestrijdingsmiddelen voor huishoudelijk gebruik (1969) *T. Soc. Geneesk.* 47, 644-652.

Flipse, L. P. Netherlands Committee for Phyto-Pharmacy; its terms of reference and its activities (1969) *TNO-Nieuws* 24, 511-514.

Fiserova-Bergerova, V., Radomski, J. L., Davies, J. E. and Davis, J. H. Levels of chlorinated hydrocarbon pesticides in human tissues (1967) *Ind. Med. Surg.* 36, 65-70.

Fitzhugh, O. G., Nelson, A. A. and Quaife, M. L. Chronic oral toxicity of aldrin and dieldrin in rats and dogs (1964) *Fd. Cosm. Toxicol.* 2, 551-562.

Fouts, J.R., Factors influencing the metabolism of drugs in liver microsomes (1963) *Annals N.Y. Acad. Sci.* 104, 875-880.

Fujimoto, J. M., Foster Eich, W. and Nichols, H. R. Enhanced sulfobromophthalein disappearance in mice pretreated with various drugs (1965) *Biochem. Pharmacol.* 14, 515-524.

Garrettson, L. K. and Curley, A. Dieldrin studies in a poisoned child (1969) *Arch. Environ. Health* 19, 814-822.

228

Genderen, H. Van – Bestrijdingsmiddelen en Volksgezondheid – in: *Op leven en dood* (1964) Centrum voor Landbouwpublicaties en landbouwdocumentatie, Wageningen.

Genderen, H. Van – The toxicology of the chlorinated hydrocarbon insecticides (1965) A progress report with particular reference to the qualitative aspects of the action in warm-blooded animals, *Mededelingen Rijksfaculteit Landbouwwetenschappen, Gent*.

Genderen, H. Van – Several types of side-effects (1969) *TNO-Nieuws* 24, 524-527.

Gilbert, D. and Golberg, L. BHT oxidase, a liver microsomal enzyme, induced by the treatment of rats with butylated hydroxytoluene (1967) *Fd. Cosm. Toxicol.* 5, 481-490.

Gillett, J. W., Chan, T. M. and Terriere, L. C. Interactions between DDT analogs and microsomal epoxidase systems (1966) *J. Agr. Food Chem.* 14, 540-545.

Golberg, L. Liver enlargement produced by drugs (1966) *Proc. European Society for the study of drug toxicity* 7, 171-184.

Goodwin, E. S., Goulden, R. and Reynolds, J. G. Rapid identification and determination of residues of chlorinated pesticides in crops by Gas-liquid Chromatography (1961) *J. Soc. Anal. Chem.* 86, 697-709.

Harris, L. E., Greenwood, D. A., Butcher, J.E., Street, J. C., Legrande Shupe, J. and Biddulph C. (1961) Report to Shell Chem. Co. on longterm oral exposure of sheep to dieldrin, May 1961, *Progress Report, Utah State University Logan, Utah*.

Hart, L. G., Adamson, R. H., Dixon, R. L and Fouts, J. R. Stimulation of hepatic microsomal drug metabolism in the newborn and fetal rabbit (1962) *J. Pharmacol. Exp. Ther.* 137, 103-106.

Hart, L. G. and Fouts, J. R. Effects of acute and chronic DDT administration on hepatic microsomal drug metabolism in the rat (1963) *Proc. Soc. Exptl. Biol. Med.* 114, 388-392.

Hart, L. G., Shultice, R. W. and Fouts, J. R. Stimulatory effects of chlordane on hepatic microsomal drug metabolism in the rat (1963) *Tox. Appl. Pharmacol.* 5, 371-386.

Hart, L. G. and Fouts, J. R. Studies of the possible mechanism by which chlordane stimulates hepatic microsomal drug metabolism in the rat (1965) *Biochem. Pharmacol.* 14, 263-272.

Hatch, T. F. Changing objectives in occupational health (1962) *Amer. Industr. Hyg. Assoc. J.* 23, 1-7.

Hatch, T. F. Significant dimensions of the dose-response relationship (1968) *Arch. Environ. Health* 16, 571-578.

Hathway, D. E. and Mallinson, A. Chemical studies in relation to convulsive conditions-Effects of Telodrin on the liberation and utilization of ammonia in the rat brain (1964) *Biochem. J.* 90, 51-60.

Hathway, D. E. The biochemistry of dieldrin and Telodrin (1965) *Arch. Environ. Health* 11, 380-388.

Hathway, D.E., Moss, J.A., Rose, J.A. and Williams, D. J. M. Transport of dieldrin from mother to blastocyst and from mother to fetus in pregnant rabbits (1967) *European J. Pharmacol.* 1, 167-175.

Hayes, W. J. Dieldrin poisoning in man (1957) *Publ. Hlth. Rep. (Wash.)* 72, 1087-1091.

Hayes, W. J. The toxicity of dieldrin to man-report of a survey (1959) *Bull. Wld. Hlth. Org.* 20, 891-912.

Hayes, W. J. *Clinical Handbook of Economic Poisons* (1963) U.S. Dept. of Health, Education and Welfare, Atlanta, Georgia.

Hayes, W. J., Dale, W. E. and Burse, V. W. Chlorinated hydrocarbon pesticides in the fat of people in New Orleans (1965) *Life Sci.* 4, 1611-1615.

Hayes, W. J. Monitoring food and people for pesticide content (1966) *Scientific aspects of pest control*, Publ. No. 1402, 314-342.

Hayes, W. J. Toxicity of pesticides to man: risk from present levels (1967) Proc. Royal Society-series B-*Biol. Sciences* No. 1007, 167, 101-127.

Hayes, W. J. and Curley, A. Storage and excretion of dieldrin and related compounds: effects of occupational exposure (1968) *Arch. Environ. Health* 16, 155-162.

Heath, D. F. and Vandekar, M. Toxicity and metabolism of dieldrin in rats (1964) *Br. J. Ind. Med.* 21, 269-279.

Heiden, J.A. Van de, and Jonge, H. De – Average length of absence during illness (1968) *CAT-rapport* 53/68, Shell Nederland Chemie (unpublished).

Hine, C. H. Results of reproduction study of rats fed diets containing endrin insecticide over three generations (1965) Unpublished report No. 2 of Hine Laboratories Inc., San Francisco.

Hine, C. H. Results of reproduction study of rats fed diets containing dieldrin over three generations (1967) *Hine Laboratories Inc. Report* No. 4, May 1967.

Hodge, H. C., Boyce, A. M., Deichmann, W. B., Kraybill, H. F. Toxicology and no-effect levels of aldrin and dieldrin (1967) *Tox. Appl. Pharmacol.* 10, 613-675.

Hoffman, W. S., Fishbein, W. I. and Andelman, M. B. The pesticide content of human fat tissue (1964) *Arch. Environ. Health* 9, 387-394.

Hoffman, W. S., Adler, H., Fishbein, W. I. and Bauer, F. C. Relation of pesticide concentrations in fat to pathological changes in tissues (1967) *Arch. Environ. Health* 15, 758-765.

Hoffman, W. S. Clinical evaluation of the effects of pesticides on man (1968) *Ind. Med. Surg.* 37, 289-292.

Hoogendam, I., Versteeg, J. P. J. and De Vlieger, M. Electroencephalograms in insecticide toxicity (1962) *Arch. Environ. Health* 4, 86-94.

Hoogendam, I., Versteeg, J. P. J. and De Vlieger, M. Nine years toxicity control in insecticide plants (1965) *Arch. Environ. Health* 10, 441-448.

Hootsmans, W. J. M. De klinische betekenis van het electroencephalogram (1962) *Ned. T. Geneesk.* 106, 2584-2591.

Hunter, C. G., Robinson, J. and Richardson, A. Chlorinated insecticide content of human body fat in Southern England (1963) *Brit. Med. J.* 1, 221-224.

Hunter, C. G., Robinson, J. and Jager, K. W. Aldrin and dieldrin-the safety of present exposures of the general population of the United Kingdom and the United States (1967) *Fd. Cosm. Toxicol.* 5 781-787.

Hunter, C. G. and Robinson, J. Pharmacodynamics of dieldrin (HEOD) Part I (1967) *Arch. Environ. Health* 15, 614-626.

Hunter, C. G. and Robinson, J. Aldrin, dieldrin and man (1968) *Fd. Cosm. Toxicol.* 6, 253-260.

Hunter, C. G. Allowable body burdens of organochlorine pesticides (1968) Paper read at the 5th Internat. Congress of Hygiene and Preventive Medicine-Rome, 8-12 Oct. 1968.

Hunter, C. G., Robinson, J. and Roberts, M. Pharmacodynamics of dieldrin (HEOD) Part II: Ingestion by human subjects for months 18-24 and post-exposure for 8 months (1969) *Arch. Environ. Health* 18, 12-21.

Ibrahim, T. M. A toxicological study of the action of the insecticide dieldrin and related substances on the contraction of striated muscle in the rat (1964), *Thesis,* Faculty of Veterinary Medicine, University of Utrecht, Elinkwijk-Utrecht-Holland.

Jacobziner, H. and Raybin, H. W. Poisoning by insecticide (endrin) (1959) *New York J. Med.* 59, 2017-2022.

Jager, K. W. and Versteeg, J. P. J. Health of workers with long-term high exposure to aldrin and dieldrin in the manufacturing plant at Pernis, Holland (1967) Appendix V Submission by Shell International Chemical Co. Ltd., to the Advisory Committee on Pesticides and Toxic Chemicals in the U.K.

Jager, K. W. and Versteeg, J. P. J. Health of workers with long-term high exposure to aldrin and dieldrin in the formulation plant at Pernis, Holland (1967) Appendix VI to above.

Jonge, H. De – Halfwaardetijd Telodrin en dieldrin gehaltes (1970) *CAT-rapport* 107/70, Shell Nederland Chemie, unpublished.

Jonge, H. De – Ziekteverzuim chemische fabrieken (1970) *CAT-rapport* 104/70, Shell Nederland Chemie, unpublished.

Kay, K. Recent research on esterase changes in mammals, by organophosphate, carbamate and chlorinated hydrocarbon pesticides (1966) *Ind. Med. Surg.* 35, 1068-1074.

Kazantzis, G., McLaughlin, A. I. G. and Prior, P. F. Poisoning in industrial workers by the insecticide aldrin (1964) *Brit. J. Ind. Med.* 21, 46-51.

Keane, W. T. and Zavon, M. R. The total body burden of dieldrin (1969-a) *Bull. Environ. Cont. Tox.* 4, 1-16.

Keane, W. T. and Zavon, M. R. Validity of a critical blood level for prevention of dieldrin intoxication (1969-b) *Arch. Environ. Health* 19, 36-44.

Keane, W. T., Zavon, M. R. and Witherup, S. H. Dieldrin poisoning in dogs: relation to obesity and treatment (1969) *Brit. J. Industr. Med.* 26, 338-341.

Kehoe, R. A. Contaminated and natural lead environments of man (1965) *Arch. Environ. Health* 11, 736-739.

Khairy, M. Effects of chronic dieldrin ingestion on the muscular efficiency of rats (1960) *Br. J. Ind. Med.* 17, 146-148.

Kiigemagi, U., Sprowls, R. G. and Terriere, L. C. Endrin content of milk and body tissues of dairy cows receiving endrin daily in their diet (1958) *J. Agric. Fd. Chem.* 6, 518-521.

Kimbrough, R. D., Gaines, T. B. and Hayes, W. J. Combined effect of DDT, pyrethrum and piperonyl butoxide on rat liver (1968) *Arch. Environ. Health.* 16, 333-341.

Korte, F., Ludwig, G., Stiasni, M.,Rechmeier, G. and Kochem, W. Metabolic studies with C^{14}-labelled Drin-insecticides (1963) Paper presented at the Vth International Pesticide Congress, London July 1963.

Korte, F. Metabolism of chlorinated insecticides (1967) Paper presented at the Scientific Plant Protection Congress, Vienna, September 1967.

230

Korte, F. Metabolism of ^{14}C-labelled insecticides in microorganisms, insects and mammals (1967) *Botyu-Kagaku* 32, 46-59.

Korte, F. Recent results in studies on the fate of chlorinated insecticides (1968) Sixth Inter-American Conference on Toxicology and Occupational Medicine, August 26-29 1968, Florida.

Kraybill, H. F. Significance of pesticide residues in foods in relation to total environmental stress (1969) *Canad. Med. Ass. J.* 100, 204-215.

Kuntzman, R., Jacobson, M. and Conney, A. H. Effect of phenylbutazone on cortisol metabolism in man (1966) *Pharmacologist* 8, 195.

Kuntzman, R., Jacobson, M., Levin, W. and Conney, A. H. Stimulatory effect of N-phenylbarbital (phetharbital) on cortisol hydroxylation in man (1968) *Biochem. Pharmac.* 17, 565-571.

Kunze, F. M. and Laug, E. P. Toxicants in tissues of rats on diets containing dieldrin, aldrin, endrin and isodrin (1953) *Fed. Proc.* 12, 339.

Kupfer, D. Effects of some pesticides and related compounds on steroid function and metabolism (1967) *Residue Reviews* 19, 11-30.

Leven met insecten (1969) Centrum voor Landbouwpublicaties, Wageningen.

Lichtenstein, E. P. Insecticides in the environment (1968) Paper read at the symposium on the science and technology of residual insecticides in food production with special reference to aldrin and dieldrin-Gaithersburg, Md. U.S.A. June 1968.

London, M. and Pallade, S. L'investigation de l'excitabilité nerveuse du rat intoxiqué par l'aldrin (1964) *Med. Lavoro* 55, 589-597.

Long, K. R., Beat, V. B., Gombart, A. K., Sheets, R. F., Hamilton, H. E., Falaballa, F., Bonderman, D. P. and Choi, U. Y. The epidemiology of pesticides in a rural area (1969) *Amer. Ind. Hyg. Assoc. J.* 30, 298-304.

Lu, F. C., Jessup, D. C. and Lavallée, A. Toxicity of pesticides in young versus adult rats (1965) *Fd. Cosm. Toxicol.* 3, 591-596.

Ludwig, G., Weis, J. and Korte, F. Excretion and distribution of aldrin ^{14}C and its metabolites after oral administration for a long period of time (1964) *Life Sci.* 3, 123-130.

Ludwig, G., Arent, H., Kochen, W., Poonawalla, N., Rechmeier, G., Stiasni, M., Vogel, J., and Korte, F. Metabolism of chlorinated insecticides by living organisms(1966)Paper presented at the Scientific Plant Protection Conference Budapest, Hungary, Febr. 22-25, 1966.

McGill, A.E.J. and Robinson, J. Organochlorine insecticide residues in complete prepared meals, a twelve month survey in S.E. England (1968) *Fd. Cosm. Toxicol.* 6, 45-57.

McGill, A. E. J., Robinson, J. and Stein, M. Residues of dieldrin (HEOD) in complete prepared meals in Great Britain during 1967 (1969) *Nature* 221, 761-762.

Michel, H. O. An electrometric method for the determination of red blood cell and plasma cholinesterase activity (1949) *J. Lab. and Clin. Med.* 34, 1564-1568.

Microsomes, metabolism and toxicity-Articles of general interest (1969) *Fd. Cosm. Toxicol.* 7, 659-662.

Mills, P. A. Pesticide residue content (1963) *Journal A.O.A.C.* 46, 762-767.

Moss, J. A. and Hathway, D. E. Transport of organic compounds in the mammal partition of dieldrin and Telodrin between the cellular components and soluble proteins of the blood (1964) *Biochem. J.* 91, 384-393.

Muir, C. M. C. Acute toxicity of aldrin, dieldrin and endrin formulations (1968) *Tunstall Report* Dec. 6th 1968 (unpublished).

Natoff, I. L. and Reiff, B. The effect of dieldrin (HEOD) on chronaxie and convulsion thresholds in rats and mice (1967) *Tunstall Laboratory Report* R (T)-3-67.

Nelson, E. Aldrin poisoning (1953) *Rocky Mtn. Med. J.* 50, 483-486.

Nelson, S. C., Bahler, T. L., Hartwell, W. V., Greenwood, D. E. and Harries, L. E. Serum alkaline phosphatase levels, weight changes and mortality rates of rats fed endrin (1956) *J. Agric. Fd. Chem.* 4, 696-700.

Novak, A. F. and Ramachandra Rao, M. R. Food safety program; Endrin monitoring in the Missisippi river (1965) *Science* 150, 1732-1733.

Oser, B. L. Much ado about safety (1969) *Fd. Cosmet. Toxicol.* 7, 415-424.

Peakall, D. B. Pesticide-induced enzyme breakdown of steroids in birds (1967) *Nature* 216, 505-506.

Pesticides-a code of conduct (1968) Alembic House, London S.E. 1.

Pincherle, G. and Shanks, J. Value of the erythrocyte sedimentation rate as a screening test (1967) *Br. J. Prev. Soc. Med.* 21, 133-136.

Popper, H. Application of histologic techniques in clinical pharmacology of the liver (1967) *Clinical Pharmacology* 8, 701-710.

Princi, F. Toxicology, diagnosis and treatment of chlorinated hydrocarbon insecticide intoxications (1957) *Arch. Industr. Health* 16, 333-336.

Quaife, M. L., Winbush, J. S. and Fitzhugh, O. G. Survey of quantitative relationships between ingestion and storage of aldrin and dieldrin in animals and man (1967) *Fd. Cosmet. Toxicol.* 5, 39-50.

Raalte, H.G.S. Van – Antagonism towards pesticides (1964) *Shell Med. Bull.* 2, 29-33.

Raalte, H.G.S. Van – Aspects of pesticide toxicity (1965) Paper presented at the Conference on Occupational Health, Caracas, Venezuela.

Raalte, H.G.S. Van – Toxicological considerations (1966) *Proc. of the Vth Shell Ind. Doct. Meeting,* 11-13 May 1966-Godorf, 124-128.

Raalte, H.G.S. Van – Legislation and the use of pesticides (1967) *Ind. Med. Surg.* 36,145-146.

Raalte, H.G.S. Van – Industrie en beperking van ongewenste beinvloeding van het milieu (1969) *Landbouwkundig tijdschrift* 81, 160-162.

Radeleff, R. D. Hazards to livestock of insecticides used in mosquito control (1956) *Mosquito News* 16, 79-80.

Radomski, J. L., Deichmann, W. B., Clizer, E. E. and Rey, A. Pesticide concentrations in the liver, brain and adipose tissue of terminal hospital patients (1968) *Fd. Cosm. Toxicol.* 6, 209-220.

Reddy, D. B., Edward, V. D., Abraham, G. J. S. and Venkaliswara Rao, K. Fatal endrin poisoning (1966) *J. Indian Med. Assoc.* 46, 121-124.

Reece, R. L. An analysis of 4000 chemistry graphs (1968) *Technicon Quarterly* 1, 7-12.

Reitman, S. and Frankel, S. Colorimetric method for the determination of serum transaminase activity (1957) *Am. J. Clin. Pathol.* 28,56-63.

Remmer, H. Die Induktion arzneimittelabbauender Enzyme im endoplasmatischen Retikulum der Leberzelle durch Pharmaka (1967) *Dtsch. Med. Wschr.* 92, 2001-2008.

Report of the Committee on persistent pesticides-1968-Division of biology and agriculture, National Research Council to U.S. Department of Agriculture. Conclusions and recommendations quoted in (1969) *T. Soc. Geneesk.* 47, 801.

Report by the Advisory Committee on pesticides and other toxic chemicals ("Wilson Committee") *Further review of certain persistent organochlorine pesticides used in Great Britain* (1969) H.M.S.O. London.

*Review of the persistent organochlorine pesticides,*Report by the Advisory Committee on Poisonous Substances used in agriculture and food storage. Chairman: J. Cook, London, H.M.S.O. 1964.

*Review of the persistent organochlorine pesticides,*Supplementary report by the Advisory Committee on Pesticides and Toxic Chemicals. Chairman: J. Cook, London, H.M.S.O. 1964.

Revzin, A. M. Effects of endrin on telencephalic function in the pigeon (1966) *Tox. Appl. Pharmacol.* 9, 75-83.

Richardson, L. A., Lane, J. R., Gardner, W. S., Peeler, J. T. and Campbell, J. E. Relationship of dietary intake to concentration of dieldrin and endrin in dogs (1967) *Bull. Environ. Contam. Toxic.* 2, 207-219.

Richardson, A., Robinson, J., Bush, B. and Davies, J. M. Determination of dieldrin (HEOD) in blood (1967) *Arch. Environ. Health* 14, 703-708.

Richardson, A., Baldwin, M. K. and Robinson, J. Metabolites of dieldrin (HEOD) in the urine and faeces of rats (1968) *Chemistry and Industry* 588-589.

Robinson, J. (1962)-Private communication Shell Research Ltd.

Robinson, J. Determination of dieldrin in blood by Gas-liquid Chromatography (1963) *Proc. 3rd. Int. Meeting in Forensic Immunology, Medicine, Pathology and Toxicology,* London, April 16-24, 1963.

Robinson, J., Richardson, A., Hunter, C. G., Crabtree, A. N. and Rees, H. J. Organochlorine insecticide content of human adipose tissue in South Eastern England (1965) *Br. J. Ind. Med.* 22, 220-229.

Robinson, J. and Mc.Gill, A. E. J. Organochlorine insecticide residues in complete prepared meals in Great Britain during 1965 (1966) *Nature* 212, 1037-1038.

Robinson, J. and Hunter, C. G. Organochlorine insecticides: concentrations in human blood and adipose tissue (1966) *Arch. Environ. Health* 13, 558-563.

Robinson, J. Dynamics of organochlorine insecticides in vertebrates and ecosystems (1967) *Nature* 215, 33-35.

Robinson, J., Richardson, A. and Brown, V. K. H. Pharmacodynamics of dieldrin in pigeons (1967) *Nature* 213, 734-736.

Robinson, J. and Roberts, M. Accumulation, distribution and elimination of organochlorine insecticides by vertebrates (1967) Paper presented at the Symposium on Physicochemical and Biophysical factors affecting the activity of pesticides, London, April 10-12, 1967.

Robinson, J. Persistent pesticides (1969) *Ann. Rev. Pharmacol.* 9, to be published.

Robinson, J. The burden of chlorinated hydrocarbon pesticides in man (1969) *Canad. Med. Ass. J.* **100**, 180-191.

Robinson, J. and Roberts, M. Estimation of the exposure of the general population to dieldrin (HEOD) (1969) *Fd. Cosm. Toxicol.* **7**, 501-514.

Robinson, J., Roberts, M., Baldwin, M. and Walker, A. I. T. The pharmacokinetics of HEOD (dieldrin) in the rat (1969) *Fd. Cosm. Toxicol.* **7**, 317-331.

Roe, F. J. C. Carcinogenesis and sanity (1968) *Fd. Cosm. Toxicol.* **6**, 485-498.

Rubin, E., Hutterer, F. and Lieber, C. S. Ethanol increases smooth endoplasmic reticulum and drug-metabolizing enzymes (1968) *Science* **159**, 1469-1470.

Schafer, M. L. Pesticides in blood (1968) *Residue Revue* **24**, 19-39.

Schalm, M. De SGOT- en SGPT-waarden in de herstelphase van acute leverparenchymbeschadiging (1969) *Ned. T. Geneesk.* **113**, 427-428.

Sherman, M. and Rosenberg, M. M. Subchronic toxicity of four chlorinated dimethanonapthalene insecticides to chicks (1954) *J. Econ. Entom.* **47**, 1082-1083.

Smyth Jr., H. F. Sufficient challenge (1967) *Fd. Cosm. Toxicol.* **5**, 51-58.

Soloway, S. Correlation between biological activity and molecular structure of cyclodiene insecticides (1965) *Advances in Pest Control Research* **6**, 85-126.

Sowell, W. L., Lawrence, C. H. and Coleman, R. L. Endrin, a review (1968) *J. Okla. State Med. Ass.* **61**, 163-169.

Speck, L. B. and Maaske, C. A. The effects of chronic and acute exposure of rats to endrin (1958) *Arch. Ind. Health* **18**, 268-272.

Spiotta, E. J. Aldrin poisoning in man (1951) *Arch. Ind. Hyg. Occ. Med.* **4**, 560-566.

Stemmer, K. L. and Hamdi, E. Electronmicroscopic changes in the livercells after prolonged feeding of DDT and Heptachlor (1966) *The Kettering Laboratory, Cincinnati, unpublished report.*

Stevenson, D. E. The toxicity of Telodrin (1964) *Mededelingen Rijksfaculteit Landbouwwetenschappen, Gent* **29**, 1198-1207.

Stevenson, D. E. The assessment of possible health hazards associated with the use of pesticides (1966) *Chemy. Ind.* 690-694.

Stevenson, D. E., Walker, A. I. T. and Ferrigan, L. W. Pharmacodynamics of dieldrin (HEOD)-4: Lifetime oral exposure of mice, I. Clinical aspects (1968) *Tunstall Laboratory report* TU/22/68.

Stoewsand, G. S. and Bourke, J. B. The influence of dietary protein on the resistance to dieldrin toxicity in the rat (1968) Paper presented at the 6th Inter-American Conference on Toxicology and Occupational Health, Miami, August 26-29 1968. Abstract in *Ind. Med. Surg.* (1968) **37**, 526.

Stokinger, H. E. The spectre of today's environmental pollution-U.S.A. brand: New perspectives from an old scout (1969) *Am. Ind. Hyg. Ass. J.* **30**, 195-217.

Street, J. C., Butcher, J. E., Raleigh, R. J. and Clanton, D. C. Tissue storage and transferal to the lamb of aldrin, dieldrin, and endrin when fed to bred ewes (1957) *Amer. Soc. Animal Prod. Western Sect. Proc.* **8** paper 46, 1-6.

Street, J. C. DDT-antagonism to dieldrin storage in adipose tissue of rats (1964) *Science* **146**, 1580-1581.

Street, J.C., Chadwick, R.W., Wang, M. and Philips, R.L. Insecticide interactions affecting residue storage in animal tissues (1966-a) *J. Agr. Food Chem.* **14**, 545-549.

Street, J. C., Wang, M. and Blau, A. D. Drug effects on dieldrin storage in rat tissue (1966-b) *Bull. Exp. Contamin. Toxicol.* **1**, 6-15.

Street, J. C. and Chadwick, R. W. Stimulation of dieldrin metabolism by DDT (1967) *Tox. Appl. Pharmacol.* **11**, 68-71.

Street, J. C., Foster, L., Mayer, M. S. and Wagstaff, D. J. Ecological significance of pesticide interactions (1969) *Ind. Med. Surg.* **38**, 91-96.

Sukhatme *Third World Food Survey* (1964) FAO Basic Study No. 11, Rome.

Tabor, E. C. Pesticides in urban atmospheres (1965) *J. Air. Poll. Contr. Assoc.* **15**, 415-418.

Tarrant, K. R. and Tatton, J. O'G. Organochlorine pesticides in rainwater in the British Isles (1968) *Nature* **219**, 725-727.

Terriere, L. C., Kiigemagi, U. and England, D. C. Endrin content of body tissues of steers, lambs, and hogs receiving endrin in their daily diet (1958) *J. Agr. Food Chem.* **6**, 516-518.

The regulations of Pesticides in the U.S.A. 1967.

Threshold Limit Values of Air-borne Contaminants (1969) Amer. Conf. of Governm. Ind. Hygienists, Ohio, Cincinnati.

Treon, J. F. and Cleveland, F. P. Toxicity of certain chlorinated hydrocarbon insecticides for laboratory animals, with special reference to aldrin and dieldrin (1955) *J. Agr. Food Chem.* **3**, 402-408.

Treon, J. F., Cleveland, F. P. and Cappel, J. Toxicity of endrin for laboratory animals (1955) *J. Agr. Food Chem.* **3**, 842-848.

Treon, J. F. The toxicology and pharmacology of endrin (1956) *Unpublished report of the Kettering Laboratory*, University of Cincinnati, Ohio.

Triet, A. J. Van and Frenkel, M. De interpretatie van een verhoogd gehalte aan alkalische fosfatase in het serum (1963) *Ned. T. Geneesk.* **107**, 1598-1603.

Triolo, A. J. and Coon, J. M. Toxicologic interactions of chlorinated hydrocarbon and organophosphate insecticides (1966) *J. Agr. Food Chem.* **14**, 549-555.

Triolo, A. J. and Coon, J. M. The protective effect of aldrin against the toxicity of organophosphate anticholinesterases (1966) *J. Pharmacol. Exptl. Therap.* **154**, 613-623.

Use of Pesticides. A report of the President's Science Advisory Committee, The White House U.S.A. 1963.

Varley, A. B. The generic inequivalence of drugs (1968) *JAMA* **206**, 1745-1748.

Vlieger, M. De, Robinson, J., Baldwin, M. K., Crabtree, A. N. and Van Dijk, M. C. The organochlorine insecticide content of human tissues (1968) *Arch. Environ. Health* **17**, 759-767.

Walker, A. I. T. Studies on the oral toxicity of dieldrin: two year oral exposure of dogs (1966) *Tunstall Laboratory Report* IRR TL/24/66.

Walker, A. I. T. Studies on the oral toxicity of dieldrin: lifetime oral exposure of dogs (1967) *Tunstall Laboratory Report* IRR TL/18/67.

Walker, A. I. T., Stevenson, D. E., Robinson, J., Ferrigan, L. W. and Roberts, M. Pharmacodynamics of dieldrin (HEOD)-3: Two year oral exposures of rats and dogs (1968) *Tunstall Laboratory Report* TL/23/68 (to be published).

Walker, A. I. T. and Stevenson, D. E. Pharmacodynamics of dieldrin (HEOD)-4: Lifetime oral exposure of mice. A summary of the results obtained from experiments currently in progress (1968) *Tunstall Laboratory Report.*

Wasserman, M., Curnow, D. M., Forte, P. N. and Groner, Y. Storage of organochlorine pesticides in the body fat of people in Western Australia (1968) *Ind. Med. Surg.* **37**, 295-300.

Weeks, D. E. Endrin food poisoning (1967) *Bull. Wld. Hlth. Org.* **37**, 499-512.

Weihe, M. Klorerede insekticider i fedtvaev fra mennesker (1966) *Ugeskrift Laeger* **128**, 881-882.

Weikel, J. H. Jr., Laug, E. P. and Tomchick, R. Ion movement across the rabbit erythrocyte membrane as affected by chlorinated insecticides (1958) *Arch. Int. Pharmacodyn.* **113**, 261-272.

Weinbren, K. and Ferrigan, L. W. Pharmacodynamics of dieldrin (HEOD)-4: Lifetime oral exposure of mice, II Hepatic changes (1968) *Tunstall Laboratory Report.*

Werk, E. E., McGee, J. and Sholiton, L. J. Effect of diphenylhydantoin on cortisol metabolism in man (1964) *J. Clin. Invest.* **43**, 1824-1835.

West, I. Biological effects of pesticides in the environment (1966) *Advances in Chem. Series* **60**, 38-53. *Interim Note.*

Wheatly, G. A. and Hardman, J. A. Indications of the presence of organochlorine insecticides in rainwater in Central England (1965) *Nature* **207**, 486-487.

Williams, S. Pesticide residues in total diet samples (1964) *Journ. A.O.A.C.* **47**, 815-821.

Williams, M. K. and Zielhuis, R. L. The concepts of lead-absorption and lead poisoning (1968) Paper read at the Conference Perm. limits inorg. lead, to be published.

Wiswesser, W. J. Pesticide problems (1965) *Arch. Environ. Health* **10**, 599-603.

Wit, S. L. Enige aspecten van de toxicologie en chemische analyse van bestrijdingsmiddelen residu's (1964) *Voeding* **25**, 609-628.

Wolfe, H. R., Durham, W. F. and Armstrong, J. F. Health hazards of the pesticides endrin and dieldrin (1963) *Arch. Environ. Health* **6**, 458-464.

Wong, D. T. and Terriere, L. C. Epoxidation of aldrin, isodrin and heptachlor by rat liver microsomes (1965) *Biochem. Pharmacol.* **14**, 375-377.

Worden, A. N. Toxicity of Telodrin (1969) *Tox. Appl. Pharmacol.* **14**, 556-573.

Wright, A. S. and Donninger, C. Liver cell changes induced by the oral administration of dieldrin and phenobarbitone to female dogs (1968) *Tunstall Laboratory Research Report* TLGR 0012.68.

Wright, A. S., Potter, D., Wooder, M. F. and Donninger, C. Liver cell changes induced by the oral administration of dieldrin, phenobarbitone and 4-amino-2,3-dimethylazobenzene to male mice (1969) *Tunstall Laboratory Research Report* TLGR 0049.69.

Wright, A. S., Greenland, R. D., Zavon, M. R. and Donninger, C. Liver cell changes following the oral administration of dieldrin to rhesus monkeys for 5.5-6 years (1969) *Tunstall Laboratory Interim Note*, unpublished.

Wright, J. W. (1970) cited from general press quotation, Febr. 7th 1970.

Zapp, J. A. Jr. The toxicologic evaluation of low-level long-term exposure in man. In *"Research in Pesticides"*, edited by C.O. Chichester, Acad. Press, New York (1965).

234

Zavon, M. R. Toxicology and pharmacology of Telodrin insecticide (compound SD 4402) (1961) *Unpublished report of the Kettering Laboratory,* University of Cincinnati, Sept. 1961.

Zavon, M. R. The toxicology and pharmacology of endrin (1961) *Unpublished data from the Kettering Laboratory,* University of Cincinnati, Ohio.

Zavon, M. R., Hine, C. H. and Parker, K. D. Chlorinated hydrocarbon insecticides in human body fat in the United States (1965) *JAMA* **193**, 837-839.

Zavon, M. R., Tye, R. and Stemmer, K. L. The effect of long continued ingestion of dieldrin on rhesus monkeys (1967) Paper presented at the Society of Toxicology Meetings, Atlanta, Georgia, March 1967.

Zielhuis, R. L. Factoren, die het ontstaan van een intoxicatie bepalen (1960) *Mens en Onderneming* **14**, 139-149.

Zielhuis, R. L. Theoretisch denkraam voor hygiënisch beleid III (1967) *T. Soc. Geneesk.* **45**, 427-432.

Zielhuis, R. L. *Vergiftigingen in en door het beroep* (1969-a) Stafleu, Leiden.

Zielhuis, R. L. Het vaststellen van aanvaardbare concentraties (1969-b) Suppl. 2 *T. Soc. Geneesk.* **47**, 1-25.

Zondag, H. A. De diagnostische betekenis van enkele bloedenzymen (1963) *Ned. T. Geneesk.* **107**, 1088-1094.